SpringerBriefs in Molecular Science

Biometals

Series Editor

Larry L. Barton

For further volumes:
http://www.springer.com/series/10046

Ranjan Chakraborty · Volkmar Braun
Klaus Hantke · Pierre Cornelis
Editors

Iron Uptake in Bacteria with Emphasis on *E. coli* and *Pseudomonas*

 Springer

Editors
Ranjan Chakraborty
Department of Health Sciences
College of Public Health
East Tennessee State University
Johnson City
USA

Volkmar Braun
Max-Planck-Institute
  for Developmental Biology
Tübingen
Germany

Klaus Hantke
IMIT
Universität Tübingen
Tübingen
Germany

Pierre Cornelis
Vrije Universiteit Brussel
Brussels
Belgium

ISSN 2212-9901
ISBN 978-94-007-6087-5   ISBN 978-94-007-6088-2 (eBook)
DOI 10.1007/978-94-007-6088-2
Springer Dordrecht Heidelberg New York London

Library of Congress Control Number: 2012955344

To my mentor and a great crystallographer
Dr. Dick van der Helm,
University of Oklahoma

Dr. Dick van der Helm
1933-2010

Ranjan Chakraborty

Preface

Iron is considered to be one of the most important biologically active metal ions because of its role in many vital metabolic reactions and in energy generation. In spite of its abundance in nature, microorganisms living either inside or outside the host body struggle to acquire iron from the environment. Outside the body the shortage is due to the majority of iron being in an insoluble ferric form. Inside the body, several iron binding proteins produced by the host create a shortage of iron available to the microorganisms. There is a fierce competition for acquiring iron among microorganisms in an aquatic or terrestrial environment. Inside the human body limiting iron availability to the pathogens is a part of host defense mechanism to restrict the pathogens' growth. The virulence of pathogenic bacteria is enhanced significantly by the presence of a number of iron acquisition systems and pathogens' ability to efficiently acquire iron from the host body. Therefore, iron acquisition systems have become important targets for novel drug design. Microorganisms, especially, bacteria have developed many innovative ways to acquire iron. Among these, the siderophore-mediated iron transport systems have been investigated thoroughly by investigators around the world. Siderophores are small organic molecules secreted by many bacteria and fungi under iron restricted conditions. Siderophores chelate iron with a high affinity and are transported back to the cell, where the iron is released for either storage or immediate use. The transport systems use a variety of proteins, which serve as transporting channels, binding proteins, or a part of energy transducing machinery. There are several pioneering scientists whose work and dedication led to the development of the field of iron-biology. However, as far as structural chemistry of siderophores and the siderophore receptor proteins is concerned, Dr. Dick van der Helm's contribution has been staggering. Dr. Dick passed away in 2010 and this book is humbly dedicated to him. It briefly describes his contributions to the knowledge of siderophore-mediated iron transport systems. The past decade has seen a tremendous amount of information gathered on the genetics, structure, and function of the components involved in not only the siderophore mediated, but also other types of iron acquisition systems. Chapter 1 discusses the structural advancements made during the past decade in the siderophore-mediated transport systems of

Escherichia coli. Chapter 2 reviews the different means of iron acquisition in bacteria, while Chap. 3 examines the iron transport systems in *Pseudomonas*.

I want to thank my co-authors Drs. Volkmar Braun, Klaus Hantke and Pierre Cornelis for their contributions. I am also thankful to Dr. Larry Barton and to Springer for giving me the opportunity to edit and publish this short book. I am greatly indebted to Dr. Sonia Ojo and Ms. Ilaria Tassistro for their constant support and patience. My sincere gratitude goes to Dr. Hans Vogel of University of Calgary, Dr. Allan Forsman, Dr. Chris Pritchett, Dr. Bert Lampson, and Mr. William Wright of Department of Health Sciences, East Tennessee State University for editorial assistance.

Johnson City, TN, USA   Ranjan Chakraborty

Contents

Chapter 1
Ferric Siderophore Transport via Outer Membrane Receptors of *Escherichia coli*: Structural Advancement and A Tribute to Dr. Dick van der Helm—an 'Ironman' of Siderophore Biology

Ranjan Chakraborty

Abstract In aerobic environments, iron exists as an insoluble ferric-oxihydroxide polymer with a concentration of soluble iron in a 10^{-16}–10^{-18} M range. In the human body iron remains complexed with iron binding proteins. Both of these conditions create iron restricted environments for the growth of microorganisms. Iron is an essential element for the growth of a majority of microorganisms since iron acts as a cofactor for several important enzymes, and cytochromes involved in energy generation. Microorganisms employ many different strategies to acquire iron, among them siderophore-mediated iron transport, as the most common one. These types of transport systems are mostly found in Gram-negative bacteria, where they consist of outer membrane proteins, periplasmic binding proteins, inner membrane transport proteins, and energy transducing inner membrane protein complex TonB-ExbB-ExbD. The crystal structures of the outer membrane receptors FepA and FhuA from *Escherichia coli* were solved in late 1990s, but to date the mechanism of transport and energy transduction is not completely known. Enormous amounts of structural and biochemical data have been published in the past decade. This chapter pays tribute to the contributions of Dr. Dick van der Helm of the University of Oklahoma to the field of siderophore biology and discusses the structural advancement of the components involved in the sidero-phore mediated transport systems in *E. coli*.

Keywords Iron · Siderophore · Transport · TonB

R. Chakraborty (✉)
Department of Health Sciences, College of Public Health,
East Tennessee State University, Johnson City, TN 37614, USA
e-mail: chakrabr@etsu.edu

R. Chakraborty et al. (eds.), *Iron Uptake in Bacteria with Emphasis on E. coli and Pseudomonas*, SpringerBriefs in Biometals,
DOI: 10.1007/978-94-007-6088-2_1, © The Author(s) 2013

1.1 Introduction

Iron is an essential nutrient for the majority of microorganisms due to its role in a number of vital biological processes, most importantly, energy generation. There are many ways in which microorganisms obtain iron from the surrounding environment. For most microorganisms, siderophore-mediated iron acquisition is perhaps the most preferable way to acquire iron. Siderophores are chemically diverse, small organic molecules produced under iron restricted conditions by both bacteria and fungi. They sequester ferric iron from the surroundings with a high affinity and transport it into the cells. The siderophore transport is facilitated by a multicomponent receptor mediated energy dependent system located in the cell walls of both Gram-positive and Gram-negative bacteria. The importance of siderophores as a growth factor in the cultures of Mycobacteria was detected as early as 1912 by Twort and Ingram (Twort and Ingram 1912). It was not until the 1950s that siderophores were structurally detected and studied by Neilands in *Ustilago sphaerogena*, a smut fungus producing an orange brown colored siderophore, now known as ferrichrome (Neilands 1952). The initial attempt to determine the crystal structure of the aluminum complex of mycobactin was made in 1949 using X-ray diffraction (Francis et al. 1949). This technique yielded only the molecular weight and unit cell dimension. The crystal structure was eventually solved by Hough and Rogers in 1974 (Hough and Rogers 1974). The crystal structures of ferrichrome and numerous other structurally related siderophores were solved in 1970s and 1980s by Dr. Dick van der Helm's group at the University of Oklahoma (van der Helm and Poling 1976; van der Helm et al. 1980; Teintze et al. 1981; van der Helm et al. 1981; Jalal et al. 1984a; Jalal et al. 1984b; Barnes et al. 1984; Barnes et al. 1985). This article reviews Dr. Dick van der Helm's contributions to the field of siderophore biology and describes the structural advances in the mechanism of siderophore-mediated iron transport via outer membrane receptors in *E. coli*.

Dr. Dick van der Helm

Dick was born in Velsen, The Netherlands on March 16, 1933 and passed away on April 28, 2010. It will be almost impossible to enumerate the contributions of Dr. Dick van der Helm to the field of crystallographic structural science within the scope of this article as his achievements spanned from solving hundreds of crystal structures of molecules ranging from anticancer compounds produced by marine organisms to most popularly, iron chelating siderophores and their receptor proteins. Therefore, the present article will only focus on his contributions to the field of siderophore-mediated iron transport.

Early Career

Dr. Dick van der Helm received his Masters equivalent degree in physics/chemistry in 1952 from the University of Amsterdam. He received his Doctoral in Chemistry in 1956 and his Doctor in Crystallography in 1960 from the University of Amsterdam. From the years 1955 to 1959, he worked as a research assistant to chemist/crystallographers, Dr. C. H. MacGillavry and Dr. L. L. Merritt, Jr. at the University of Amsterdam and Indiana University, respectively. According to Dick, it was Caroline MacGillavry who inspired him to become interested in molecular structure determination. He was very proud of his work as a postdoctoral research associate in the laboratory of Dr. A. L. Patterson at the Institute for Cancer Research, Philadelphia. He had a black and white photograph in his office at the University of Oklahoma, of Dr. Patterson giving instruction to his research group; pointing at that photograph, he often talked about the fond memories and the experience of working with Dr. Patterson. In 1962, Dick joined the department of Chemistry and Biochemistry at the University of Oklahoma, Norman as an assistant professor and remained there until he retired in 2002. He writes in his autobiography at the University of Oklahoma (which is posted on its website) that he came to the university with just two publications to his credit and in 40 years of his illustrious career at OU, he published 325+ research articles in peer-reviewed journals and solved the crystal structures of enumerable biologically important molecules!

Introduction to the Field of Iron Transport

Dick did not begin his research career solving the crystal structures of siderophore but his curiosity to know more about the metal chelating compounds and peptides was evident by his doctoral research on determining the crystal structures of Rhodanine and the iron chelate cupferron and also from the articles he published early in his career. Very few crystallographers know that as a part of his doctoral thesis, he also wrote the computer programs for the IBM 650 computer used for the determination and refinement of crystal structures. He later wrote similar programs for the IBM 1620 while working in Dr. Patterson's laboratory. These programs were made freely available and were used worldwide for solving crystal structures. Worldwide use of these programs brought him international recognition among crystallographers.

He was instrumental in establishing a crystallographic facility at the University of Oklahoma. In the late 1990s, he renovated the crystallographic facility at the University of Oklahoma along with the help of Dr. Ann West, and installed a rotating anode X-ray generator for macromolecular crystallography. He received an NIH career award spanning the years 1969–1974, which led him to initiate structural studies on iron chelating and transporting siderophores. In 1973, he received his first NIH grant to work on siderophores and iron transport compounds in bacteria and fungi. He continued to have this grant for 29 years. Along with the help of postdoctoral fellows, Dr. M. A. F. Jalal, Dr. Bilayet Hossain and several

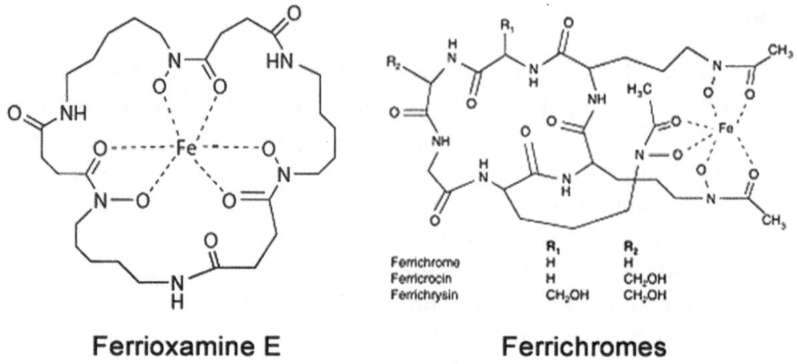

Ferrioxamine E   **Ferrichromes**

Fig. 1.1 Ferrioxamine E and ferrichromes

others who were trained by Dick, the group solved the crystal structures of several important siderophores from bacteria and fungi. These included the structures of ferrioxamine E, ferrichrome, ferrichrome A, allumichrome A, triacetyl fusarinine, ferricrocin, ferrirubin, pseudobactin, coprogens, and many more (Fig. 1.1). As Dick has mentioned in his autobiography, each structure deserved a celebration. So much so that the ceiling tiles in the laboratory got damaged by the champagne corks! The success in solving the enumerable crystal structures of siderophores led him to become interested in the mechanism of the siderophore-mediated iron transport in bacteria.

Protein crystallography and earlier work on protein purification

Dick took a sabbatical with Dr. Guenter Winkelmann at Tubingen, Germany in 1984, so that he could pursue his interest in the mechanism of siderophore-mediated iron transport. It was during this sabbatical that he interacted with renowned siderophore researchers Volkmar Braun and Klaus Hantke. These interactions and discussions culminated into an idea that the structures of siderophore transport proteins could be solved. The interest in solving the structures of outer membrane proteins involved in iron transport coupled with the confidence that these structures could be solved were greatly enhanced by the work on bacterial outer membrane hydrophilic channels called porins by Michael Garavito and Rosenbusch and a number of articles published by them during mid 1980s.

FepA structure

In the mid 1980s, Dick's group started extracting and purifying the outer membrane receptor protein FepA, which is involved in the transport of ferric-enterobactin in *E. coli*. In 1989, van der Helm and Jalal had published their first article on purification and crystallization of FepA (Jalal and van der Helm 1989). In early 1994, Dick had established collaborations with Dr. Deisenhoefer's group at the University of Texas Southwestern Medical Center in Dallas. I came to

Dick's laboratory on sabbatical in June 1994 and started working on improving the purification and crystallization of FepA along with several other outer membrane proteins involved in iron transport from *E. coli* and *Pseudomonas*. Dr. Susan Buchanan from Deisenhoefer's group got involved in the FepA project during the mid 1990s and the crystal structure of FepA was finally solved in 1998. This was perhaps one of the most satisfying and exciting moments in Dick's career. The FepA structure was solved without ligand (Buchanan et al. 1999). In spite of applying all possible methodologies and variations in crystallization conditions, FepA could not be crystallized with a bound ligand. As previously mentioned, during this period, I was also working on purifying and crystallizing other membrane proteins and had succeeded in crystallizing and obtaining some preliminary data for FhuA. Unfortunately, the structures were published independently by two other groups during the same time that we published the FepA structure. This was indeed a setback for Dick; nevertheless, he was happy about solving the FepA structure. Once the crystal structure was known, it unexpectedly revealed the presence of a plug domain occluding the passage through the beta barrel. Following the solving of the structure of FepA, we began site directed mutagenesis studies to investigate the mechanism of ferric enterobactin transport through FepA.

FecA structure

Subsequent to the success of purification of FepA, Dick's group published articles on the purification of FecA, an outer membrane receptor involved in ferric dicitrate transport in *E. coli*. Once we solved the FepA structure, I became involved in crystallizing FecA. The FecA crystals never looked beautiful or promising but to our surprise diffracted very well. The crystal structure of FecA was finally solved in 2002 with the help of crystallographer Andrew Ferguson, a member of Dr. Deisenhoefer's group. The FecA structure was solved with and without a bound ligand and the structures helped to establish the concept of bipartite gating (Ferguson et al. 2002; van der Helm et al. 2002).

Personal Side of his Character and his Philosophy on Science Education

Dick was always interested in learning about newly developed crystallographic methods. He used to work for long hours and analyzed the experimental data very carefully. It was almost always impossible to find mistakes in his calculations or interpretations. As one of his former students, Michael B. Lawson said, "Dr. van der Helm was truly a creative, pioneering scientist who would work tirelessly to accomplish (his goals). Driven to seek perfection, he opened my eyes to the marvels and wonders of the world we live in. Without the association with him I might never have come to realize the amazing and bizarre universe or multiverses that surround us. I think his work and scientific spirit can be held in the highest esteem."

Dr. van der Helm was always very careful in drawing conclusions from the experimental data and was hesitant in extrapolating his theoretical interpretations or extrapolations too far. He was a great teacher himself and always advocated for undergraduate education. He valued undergraduate education very much and later in his career he was concerned about the declining quality of undergraduate education in the nation. It is very appropriate that one of his former undergraduate research students, John Burks and his wife, have initiated an undergraduate research scholarship fund in the honor of Dick van der Helm at the University of Oklahoma. Dr. van der Helm was a great mentor to me and his services to the field of structural biology will always be remembered and missed!

1.2 Siderophore-Mediated Iron Transport Systems

A microorganism living in an aerobic environment struggles to acquire iron from the environment as the concentration of soluble iron is extremely low. In the presence of oxygen at physiological pH, ferric ions form polymers of iron oxy-hydroxide with a concentration of soluble iron in the range of 10^{-17} to 10^{-18} M (Raymond et al. 2003). Thus, soluble iron cannot get into the cell via diffusion because the concentration of iron in the cytoplasm of metabolically active bacteria is about 10^{-6} M. Within the human body, pathogens face similar shortages of iron due to the presence of iron chelating proteins like lactoferrin and transferrin, which are found in body fluids. The intracellular iron storage protein ferritin and peptide hormone hepcidin, found in the liver, help to maintain the concentration of available iron at an extremely low level (Abergel et al. 2006, 2008; Chu et al. 2010). Hepcidin, a peptide hormone synthesized by the liver binds to the iron channel ferroportin and prevents the release of iron from the gut as well as from macrophages. As a matter of fact, these strategies are part of body's immune system to restrict the growth of invading pathogens. Microorganisms use a variety of methods to acquire iron from the environment or from the human body. This is because, with a few exceptions, iron is a vital element required by almost all living cells. Iron is used as a cofactor by heme containing compounds, e.g. cytochromes and enzymes, nitrogenase, and ribonucleotide reductase (Neilands 1981). The methods employed by various bacteria to acquire iron are reviewed in the next chapter, which range from simply producing iron binding proteins on the surface of bacteria to multicomponent iron transport systems. One such method, which is widely distributed among bacteria, is siderophore-mediated iron transport systems. Siderophores have specific affinity for ferric ions. They chelate ferric ions and transport them across the membranes into the cytoplasm via energy dependent, multicomponent transport systems (Chu et al. 2010; Braun et al. 1998; Braun and Braun 2002a). The components of such systems include a receptor protein located in the outer membrane (OM), a periplasmic binding protein, and an ABC type transporter located in the inner membrane. Since there are no energy sources available in the outer membrane, the energy required for the transport via OM receptor is presumably provided by the interaction of the OM receptor with the TonB-ExbB-ExbD protein complex (will be discussed in the later sections) located in the inner membrane. This complex couples the proton motive force (PMF) created by the inner membrane (Postle 1993; Bradbeer 1993; Skare et al. 1993; Braun 1995). All TonB-dependent outer membrane transporter (TBDTs) proteins have an 8–10 residue conserved regions at their N-terminus called the TonB box, which interacts with TonB. This interaction presumably results in PMF-dependent energy transduction (Pawelek et al. 2006; Shultis et al. 2006). In addition, some of these systems have additional components involved in the positive regulation and signal transduction; this includes the ferric dicitrate transport system in *E. coli* and pyoverdine transport in *Pseudomonas* (Braun and Braun 2002b; Braun et al. 2003; Braun 1997; Cobessi et al. 2005a, b). The Gram-negative bacterial outer membranes are known to express a wide range of such TBDTs involved in not only the transport of ferric-siderophore complexes but also in the transport of antibiotics, vitamins, nickel complexes, and carbohydrates (Noinaj et al. 2010). Noinaj et al. have recently reviewed in detail the structure, function, and regulation of TBDTs (Noinaj et al. 2010). The present chapter reviews the structural advances made in the siderophore-mediated iron transport systems in *E. coli* with a focus on the mechanism of transport via OM receptor. The two other following chapters will review the different ways in which microbes acquire iron (Drs. Braun and Hantke) and the iron transport systems in *Pseudomonas* (Dr. Cornelis).

1.2.1 Ferric-Siderophore Transport Systems in E. coli

The ferric-siderophore transport systems have been thoroughly investigated in Gram-negative bacteria mostly using *E. coli* as a tool. Siderophores are structurally and chemically a very diverse group of compounds produced by many bacteria and fungi under iron restricted conditions (Chu et al. 2010; Braun et al. 1998; Braun and Braun 2002a). Siderophores chelate ferric ions with high affinity and are actively transported across the membranes. Siderophores use a variety of functional groups like catechol, hydroxamate, or carboxylate or a combination of them to chelate ferric ion. Enterobactin and aerobactin are the only endogenous siderophores reported to be produced by *E. coli*; however, *E. coli* has been reported to express seven different iron-transporting systems under iron-stressed conditions (Fig. 1.2; Postle 1993; Hantke 1984; de Lorenzo et al. 1988). The outer membrane receptors involved in these transport systems include FepA, FhuA, FecA, FhuE, IutA, Fiu, and Cir transporting ferric enterobactin, ferrichrome, ferric dicitrate, ferricoprogen, ferric aerobactin, 2,3 Dihydroxybenzoylserin (2.3-DBS), and 2,3 Dihydroxybenzoic acid (2,3 DHB), respectively (Postle 1993; Hantke 1984; de Lorenzo et al. 1988). 2,3 DBS and 2,3 DHB are the degradative products of enterobactin. Most of the receptors are multifunctional and can also serve as receptors for phages, colicins, and antibiotics (Postle 1993; Braun 1995; Killmann et al. 1995). For example, FhuA,

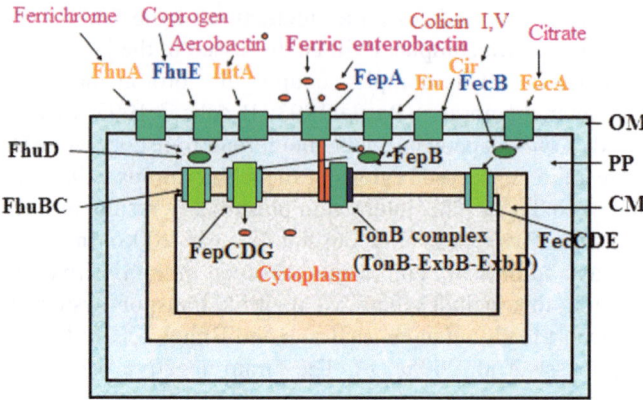

Fig. 1.2 The siderophore-mediated iron-transport systems in *E. coli*

which is primarily a ferrichrome transporting OM receptor also serves as a receptor for bacteriophages T1, T5, π80, UC-1, and toxins such as colicin M and microcin 25. The majority of the structural and biochemical advances in this area have been made using the studies on iron-transport systems of *E. coli*. The structural advances will be discussed in the following sections.

1.3 Current Structural Advancement in the Components of the Ferric-Siderophore Transport Systems from *E. coli*

Significant knowledge has been gathered in the past two decades on the structures and interactions of the components involved in the siderophore transport systems. The first reported crystal structures of TBDTs were also from *E. coli*; FhuA, and FepA were solved in late 1990s (Buchanan et al. 1999; Ferguson et al. 1998; Locher et al. 1998) followed by the liganded and unliganded structures of FecA in 2002 (Ferguson et al. 2002). The FhuA structure was solved both in unliganded and liganded forms while the structure of FepA could only be solved in its unliganded form (Buchanan et al. 1999). The crystal structures of FpvA (a ferric-pyoverdine) and FptA (a ferric-pyochelin) receptors from *Pseudomonas aeruginosa* were subsequently solved in 2005 (Cobessi et al. 2005a, b). Several articles have been published describing and analyzing these structures in light of the mechanism of transport of ferric siderophores, bacteriophages, and toxins (van der Helm and Chakraborty 2002; van der Helm 2004; Noinaj et al. 2010; Krewulak and Vogel 2008). The overall structures of TBDTs will be described in brief in the following section. This will be followed by a discussion of the structures of the periplasmic and innermembrane components.

1.3.1 Crystal Structures of TBDTs from **E. coli**

The crystal structures of FepA, FhuA, and FecA, show remarkable similarity (Fig. 1.3; Buchanan et al. 1999; Ferguson et al. 1998; Locher et al. 1998; Ferguson et al. 2002). BtuB, the TonB-dependent cobalamin transporter from *E. coli* as well as FpvA and FptA also share a common structure with these receptors (Chimento et al. 2003; Cobessi et al. 2005a, b). All of them consist of a barrel made up of 22 anti-parallel β strands and a plug domain formed by approximately 150 N-terminal residues. The extracellular loops extend about 30–40 A° above the outer leaflet of the outer membrane and form a pocket to receive the ferric siderophore complex from the external environment. Many amino acid residues lining the external loops form the initial binding site (Ferguson et al. 1998; 2002; Locher et al. 1998; Newton et al. 1999; Barnard et al. 2001; Cobessi et al. 2005a, b). From here, the ferric-siderophore complexes move further into the pocket. This binding site is formed by the residues contributed by the extracellular loops, as well as the loops formed by the plug domain. Except FepA, the structures of four receptors have been solved with bound ligand, revealing the residues involved in their binding sites. The plug domain consists of multiple short beta strands connected via loops and a few alpha helices. The plug domain is located within the barrel in such a way that it blocks access to the periplasm. The plug, formed by about 150 N-terminal residues consists of a four-stranded mixed beta sheet inclined by about 45° with respect to the membrane plane. The loops formed by the plug domain also extend high above the outer leaflet of the outer membrane and provide appropriate amino acid residues to form a second binding site (the site formed by the external loops being the first). In FecA, closing of the extracellular loops has been observed after the ligand binds to this site (Ferguson et al. 2002). These changes have not been

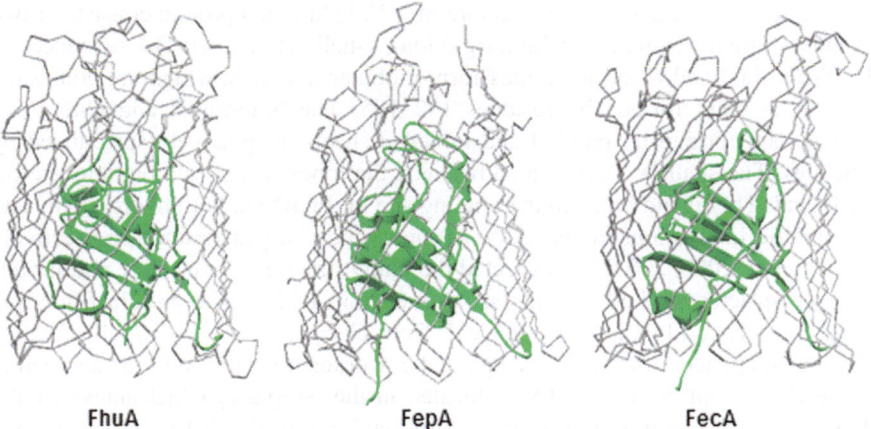

FhuA   FepA   FecA

Fig. 1.3 The crystal structures of outer membrane receptors from *E. coli*. The barrel is shown in *gray* backbone and the plug domain is depicted in *green* ribbon structure (PDB codes for FepA, FhuA and FecA are 1FEP, 1BY5 and 1KMO respectively)

observed in the case of ligand-bound FhuA structures (Ferguson et al. 1998; Locher et al. 1998), possibly due to crystal packing or another solid-state effect. Binding of ligand to the second site also induces considerable conformational changes in the plug domain. These changes extend to the periplasmic side of the membrane, extending the N-terminus into the periplasm. The first structures of the TBDTs surprised everyone by revealing a presence of a plug formed by a folded N-terminal domain occluding the passage. In a ligand-bound structure, the extension of the N-terminal domain containing the TonB box into the periplasm indicated a significant role of TonB interaction in the ligand transport. It was obvious that TonB box plays a crucial role in opening of the pore either via dislodgement or rearrangement of the plug. The structures of C-terminal TonB of various lengths were solved by both crystallography and NMR studies (Chang et al. 2001; Peacock et al. 2005). This was followed by recent reports of the structures of FhuA in complex with the periplasmic domain of TonB as well as the structure of an extended N-terminus of FecA (Pawelek et al. 2006; Garcia-Herrero and Vogel 2005).

1.3.2 Periplasmic Binding Protein Connects the Transport Process via OM and CM

The ferric-siderophore molecules are received by a periplasmic binding protein (PBP) upon their passage from the OM receptors. FhuD is a prototype of such a PBP involved in the transport of hydroxamate-siderophores belonging to the ferrichrome family. FhuD is also involved in the transport of the antibiotic albomycin (Clarke et al. 2002). The crystal structures of FhuD were recently solved in complex with a ferrichrome analog gallichrome, desferal and antibiotic albomycin (Clarke et al. 2000, 2002). The structure of a 32 kDa FhuD protein consists of two globular domains connected with a rigid long α-helix (Fig. 1.4). The sequences of the N-terminal and C-terminal ends form globular domains, which are connected with an α- helix formed by residues 142–165. The N-terminal and C-terminal domains are formed by parallel and mixed β sheets, respectively surrounded by α-helices. The shallow cleft formed between the lobes serves as a binding site for the ligand. The ligand binds to the binding site on FhuD via both hydrophobic and hydrophilic interactions between the aromatic and charged residues, respectively (Clarke et al. 2000, 2002). As expected the FhuD structure shares close similarities with the structures of other PBPs such as vitamin-B12 uptake protein BtuF, and ferric enterobactin binding protein CeuE from *Campylobacter jejuni*, all of which are TonB-dependent transports. Vogel and coworkers recently reported a structure of the C-terminal domain of ExbD located in the periplasm which unexpectedly showed very close similarities to the C-terminal lobes of the PBPs (Garcia-Herrero et al. 2007). The primary function of FhuD protein is to bind ligand and deliver it to the ABC type inner membrane permease. Recently, it has been reported that TonB acts as a scaffold for FhuD facilitating the exchange of ferric-siderophore

Fig. 1.4 The crystal
structure of FhuD bound to
gallichrome; α helices are
shown in *green* and *yellow*
while β-strands are in *blue*
and *red* ribbons. The bound
gallichrome molecule is
depicted in sticks (PDB
code1EFD)

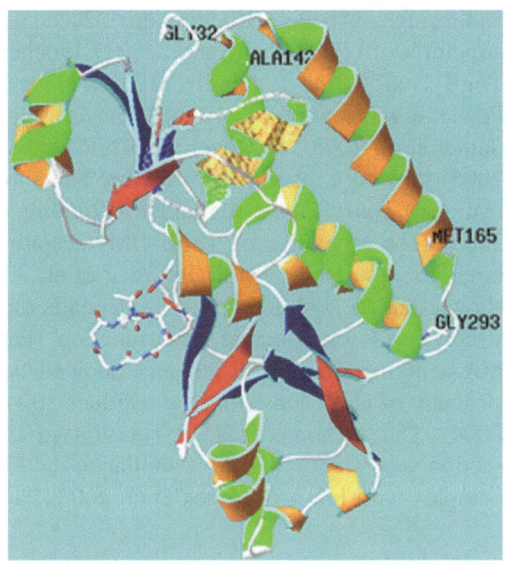

between the OM receptor and FhuD ligand binding site (Carter et al. 2006). This
seems to be a plausible model considering a relatively low affinity that exists
between the ligand and the FhuD (Clarke et al. 2000, 2002; Carter et al. 2006).
The presence of TonB acting as a scaffold may increase the efficiency of the
transfer of the ligand to FhuD.

1.3.3 Crystal Structures of TonB and its Complex with FhuA

TonB is anchored to the inner membrane by means of a single transmembrane
domain. The C-terminus is located in the periplasmic region. Structurally, TonB is
divided into three distinct domains. The amino terminal residues 1–33 are anchored
to the inner membrane and are responsible for its interaction with ExbB-ExbD. The
second domain consists of the residues 34–154 with Pro-Glu and Pro-Lys repeats.
The third C-terminal domain (CTD) consists of the remaining residues 155–239.
Domains 2 and 3 are located in the periplasm. The third domain can contact the
N-terminus of the ferric-siderophore OM receptors (Pawelek et al. 2006; Shultis
et al. 2006). As previously mentioned, TonB-dependent outer membrane receptor
proteins have 8–10 conserved residues at their N-terminus, known as the TonB box,
which interacts with the C-terminus of TonB (Lundrigan and Kadner 1986;
Gudmundsdottir et al. 1989; Sauer et al. 1990). Unfortunately, the attempts to
isolate and purify the whole TonB protein have been unsuccessful due to its
instability. However, recently, several groups have determined structures of TonB
with variable numbers of C-terminal residues. These structures include carboxyl
terminal residues 155–239 as a tightly intertwined dimer (Chang et al. 2001)

and residues 148–239 (Ködding et al. 2005) as a loose dimer. Based on the dimeric structures of CTD (Chang et al. 2001), biochemical and genetic experiments using TonB-ToxR fusion proteins by Sauer et al., and cysteine mutagenesis by Ghosh and Postle, it was reported that TonB acts as a dimer in vivo (Ghosh and Postle, 2005; Sauter et al. 2003). However, an NMR study of residues 103–239 (Peacock et al. 2005) showed a monomeric structure and subsequently, it was shown that dimerization of TonB is not essential for its binding to OM receptor (Ködding et al. 2004, 2005). The structures of TonB-OM receptor complex showed that the receptor binds to monomeric TonB (Pawelek et al. 2006, Shultis et al. 2006). It was also shown that their oligomeric state depends on the length of the recombinant constructs used for structure determination (Pawelek et al. 2006) since the CTD fragments of 77 and 86 residues showed homodimerization while the solution structure of CTD consisting the residues 103–239 was a monomer (Ködding et al. 2004). The monomeric NMR solution structure of CTD described in the next section was structurally most similar to the CTD structure subsequently solved in complex with FhuA (Peacock et al. 2005; Pawelek et al. 2006).

1.3.3.1 The Solution Structure of CTD

In the solution structure of CTD (residues 103–239), the residues 103–151 were not ordered while the residues 152–239 formed a well-ordered structure consist of two α helices and four β strands (Fig. 1.5; Peacock et al. 2005). The NMR structure included the residue Gln160 shown to be important for TBDT-TonB interaction (Vakharia-Rao et al. 2007). The oblong shaped structure of CTD with a dimensions of $45 \times 26 \times 27$ Å showed slight curvature along the long axis generating convex and concave sides (Peacock et al., 2005). The structured domain contained residues 152–164 on the concave side while the residues 165–169 formed a short α-helix. The residues 174–182 and 188–196 formed β1 and β2 strands. respectively. The residues 203–211 formed a α-helix 2 while the residues 221–230 and 236–237 formed β3 and a short β4 strand, respectively (Peacock et al. 2005). The solution structure of CTD 103–239 showed both similarity and significant differences with the dimeric crystal structures of TonB-CTD. The residues 170–195 and 199–231 of the monomeric structure overlay well with the central portion of the–crystal structure consisting of chain A and B, respectively (Chang et al. 2001; Ködding et al. 2005). The differences are observed both at the C and N terminals (Peacock et al. 2005; Chang et al. 2001; Ködding et al. 2005). In a dimeric structure, the residues 232–237 a part of β3 strand of one TonB molecule hydrogen bonds with residues 168–172 on the β3 and residues 227–231 of the β3 of the another TonB molecule. The same residues in a monomeric structure are part of a turning conformation and form a short antiparallel fourth β strand with respect to β3 strand. The residues 165–170, a part of β strand in a dimeric structure form an α-helical conformation in a monomeric structure. Even

Fig. 1.5 The NMR solution structure of C-terminal TonB residues 151-239. Four β-strands are shown in *yellow* color while the two α-helices are shown in *magenta* color (PDB code 1XX3)

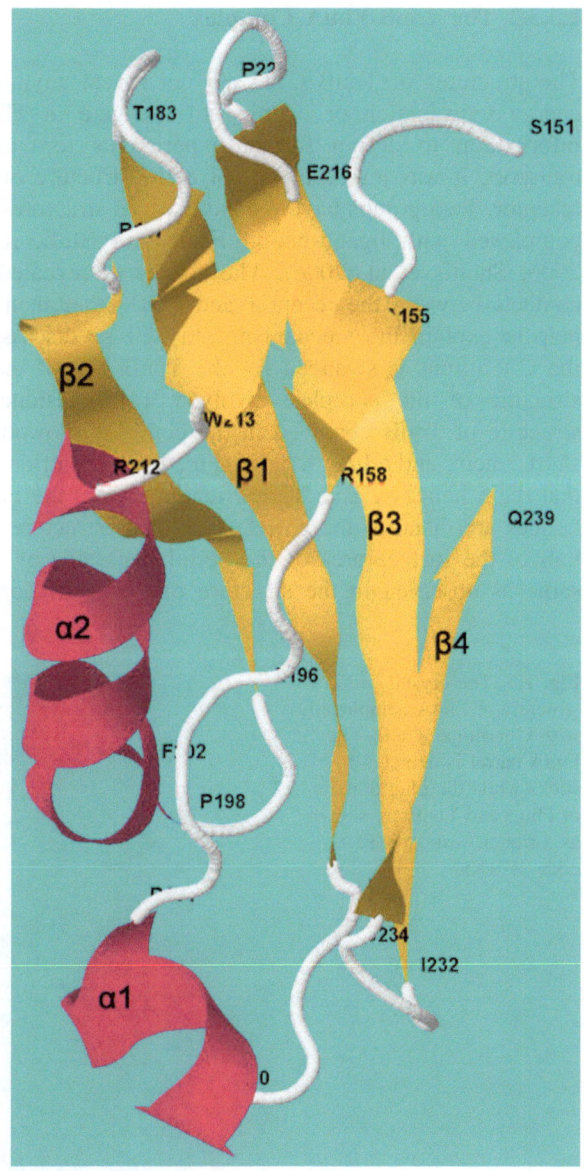

though a part of the monomeric structure overlays well with the dimeric structure, the secondary structural components in both are different. Usually in domain-swapped proteins, the secondary structures are swapped but are not changed, as is observed in this case (Rousseau et al. 2003; Peacock et al. 2005).

1.3.3.2 The TonB-FhuA Complex

The structures of TonB CTD did not reveal anything about the mechanism of
energy transduction or opening of the gate in TBDT but certainly provided
information to design further experiments to understand the mechanism of
transport. It was important to solve the structure of TonB in complex with the
receptor. Two groups have recently solved structures of the C-terminus of TonB,
complexed with ligand-bound FhuA and BtuB, respectively (Pawelek et al.,
2006; Shultis et al. 2006). The structures revealed some additional points of
contacts between the receptor and CTD other than the TonB box, which may
help to explain the mechanism and will be discussed later. In both structures,
the C-terminus is swapped for the TonB box of the OM receptor protein. The
structure of the complex involved TonB residues 158–235 (Fig. 1.6). The
structure of TonB observed in the complex showed a three-stranded β sheet, a
short α helix and a long α helix (Fig. 1.6). The orientation of the complex is such
that the α helices 1 and 2 are oriented toward the FhuA barrel while the three β
helices are distal to the barrel. The TonB structure occupies approximately one
half of the surface area of the periplasmic side of the FhuA barrel. This is the
same as observed in the structure of the BtuB-TonB complex (Pawelek et al.

Fig. 1.6 The crystal
structure of FhuA complexed
with C-terminal TonB. The
FhuA barrel is depicted in
sticks while the plug domain
of FhuA and TonB are shown
as a ribbon structure (PDB
code 2GRX)

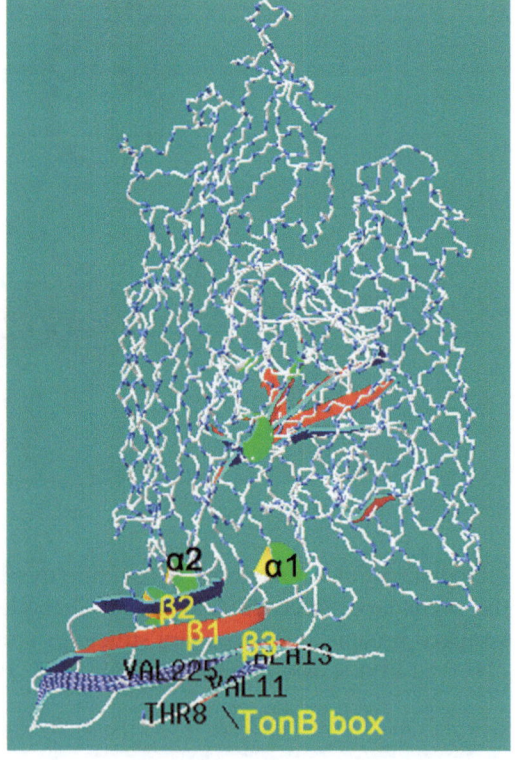

2006; Shultis et al. 2006). The positioning of the bound TonB occludes the barrel lumen from the edge containing the periplasmic loops 8–10. The interaction, however, does not result in a concerted change in the structures of the OM receptor proteins besides those which are involved directly with the motion of TonB box. The structural changes in FhuA upon the binding of TonB were minor and involved the periplasmic turns of the barrel and the periplasm exposed loop of the cork domain (Pawelek et al. 2006). On the other hand, the TonB structure showed considerable changes upon binding to FhuA. In TonB, the residue Ile232, which connects to the short $\beta 4$ in a turning conformation and the residue Thr235 on a short β strand move at the most by 2.0 and 3.5 Å, respectively. These movements are justifiable in the structure as it facilitates the parallel β interaction of the TonB box with $\beta 3$ strand of TonB (Pawelek et al. 2006; Shultis et al. 2006). The interaction involves the residues Val225, Val226, Leu229, and Lys231 of the C-terminal 'TonB' and the residues Ile9, Thr10, Val11, and Ala13 of TonB box on the FhuA plug domain (Fig. 1.6). Another interesting interaction involved the residue Gln160, which was thought to be interacting with TonB box based on mutational and biochemical analyses (Vakharia-Rao et al. 2007). The residue Gln160 in TonB possibly interacts via hydrogen bond formation with Thr12 inTonB box of FhuA (Pawelek et al. 2006). Even though the TonB box is well conserved among TBDTs, the interactions between the TonB box and the receptors FhuA and BtuB were not found to be identical (Pawelek et al. 2006; Shultis et al. 2006). In TonB-BtuB complex only the even number of residues on TonB box were involved in hydrogen bond formation with the TonB while in case on FhuA-TonB interaction, residues Ile9, Thr10, Val11, and Ala13 were involved in hydrogen bond formation with TonB. The short fourth antiparallel β strand formed by the residues 236–237 in the NMR structure of the monomeric TonB was replaced by the TonB box, which formed a parallel β strand to $\beta 3$ strand of TonB (Peacock et al. 2005; Pawelek et al. 2006; Shultis et al. 2006). Thus, the FhuA-TonB interaction resulted in β-strand exchange. In addition to the interactions between the residues located on TonB box of FhuA, nine more hydrophilic interactions were observed between the residues located on the plug and barrel domain with the TonB protein. The residues Ala26 and Glu56 of the plug domain interacted with the residue Arg166 of $\alpha 1$ region of TonB (Pawelek et al. 2006). Ala591 and Asn594 of the FhuA barrel also interacted with the residue Arg166 on TonB. The residues, Asp676, and Gly684 interacted with the residue Asn200 located on $\alpha 2$ helix of TonB while Arg724 and Phe725 showed interaction with the residue Arg204 on $\alpha 2$ helix of TonB (Pawelek et al. 2006). Do the same residues of TonB interact in a similar way in BtuB-TonB complex? The structure of FhuA-TonB does not superimpose on that of the BtuB-TonB and the additional hydrogen bonding involving the cork and barrel residues observed in FhuA-TonB complex is not completely conserved in BtuB-TonB complex. However, the same residues in TonB do form hydrogen bonds with the residues which are to some extent structurally conserved. This is not unexpected since the mature polypeptide of BtuB is much shorter (594 residues vs. 725 residues of FhuA) than FhuA, even though they share a similar domain structure. Arg166 of TonB interacts with Val22 on the cork domain and Glu423 located near the periplasmic turn of the $\beta14$ strand in a way that its side chain points to the barrel lumen facilitating its interaction with TonB. Asn200 of TonB interacts with Gly382 located on the periplasmic turn connecting β strands 12 and 13 of the BtuB barrel. The residue Arg204 of TonB interacts with Asp471 located near the periplasmic turn of the $\beta16$ strand of the barrel in a way that its side chain points to the lumen of the BtuB barrel. The interactions may not be conserved completely structurally or genetically but in both the cases the orientation of the plug and barrel residues involved in interaction with the TonB residues is similar in both complexes signifying their possible role either in additional signaling about the ligand-occupancy status of TBDTs or in pore formation. These interactions do bring about significant changes in the plug domain; however, these changes do not give any clue to the possible mechanism. Also, the structures of the complexes do not resolve the manner in which the TonB complex transduces the energy to the OM receptor resulting in either conformational changes into the plug domain or dislodgement of the plug domain allowing the transport of the ferric-siderophore complex.

1.4 Roles of ExbB and ExbD

The energy transduction by TonB is dependent on two other inner membrane proteins ExbB and ExbD, which form a complex with TonB. The ratio of TonB:ExbB:ExbD in cells is 1:7:2 (Higgs et al. 2002; Pramanik et al. 2011) but the ratio while actively transducing energy is not really known. In the complex, ExbB has a distinctly different structural topology than TonB and ExbD. ExbB polypeptide traverses the membrane thrice and has a significant portion of the C-terminal domain in the cytoplasm. The 244 residue long ExbB polypeptide includes three transmembrane helices consisting of residues 16–39, 128–155, and 162–194 respectively, embedded in the membrane while N-terminal residues 1–15, and the residues 156–161 connecting the second and third transmembrane helices are located in the periplasm (Kampfenkel and Braun 1993). Almost 60 % of the ExbB polypeptide including the residues 40–127 connecting the first and the second transmembrane helices and the C-terminal residues 195–244 is located in the cytoplasm (Kampfenkel and Braun 1993). ExbB acts as a scaffold which holds both TonB and ExbD proteins (Pramanik et al. 2010, 2011). Both ExbD and TonB share a similar surface topology, where they are anchored with a single N-terminal transmembrane domain to the inner membrane with a large C-terminal domain located in the periplasm. As previously mentioned, Vogel and coworkers have published a solution structure of the periplasmic domain of ExbD (residues 44–141), which unexpectedly shows structural homology to the periplasmic siderophore-binding proteins like FhuD (Garcia-Herrero et al. 2007).

1.4.1 ExbD Structure

The monomeric structure of ExbD was divided into three distinct regions, namely, an N-terminal flexible tail (residues 44–63), a folded globular domain (residues 64–133), and a C-terminal short flexible region (residues 134–141) (Fig. 1.7). The folded middle region consists of two α-helices, both on the same side of the β-sheet formed by mixed five β-strands. In the center of the β-sheet is the β-strand 1, which is flanked by antiparallel β2 on one side and parallel β4 on the other side (Garcia-Herrero et al. 2007). β3 and β5 are located on the edges of the β-sheet where β3 is antiparallel to β2 while β5 is parallel to β4 strand. As mentioned earlier, the structure shows a close structural homology to FhuD, which is involved in the transport of hydroxamate-siderophores belonging to the ferrichrome family, and the antibiotic albomycin. It also shares homology with the structures of vitamin-B12 uptake protein BtuF, and ferric enterobactin binding protein CeuE from *Campylobacter jejuni*. All of these are PBPs involved in TonB-dependent transports (Garcia-Herrero et al. 2007; Clarke et al. 2000; Borths et al. 2002). Several biochemical studies published by Postle and coworkers indicated that TonB-ExbB-ExbD complex is not only held together by the inner membrane but also interacts in the periplasm to form a complex (Jana et al. 2011; Ollis and Postle 2011; Ollis and Postle 2012a, b; Ollis et al. 2012). Vogel and co-workers used two synthetic peptides referred as Glu-Pro and Lys-Pro from the proline-rich regions of the periplasmic domain of TonB consisting of residues 70–83 and 84–102

Fig. 1.7 The solution structure of the periplasmic domain of ExbD (residues 44–141) solved by multidimensional NMR (PDB code 2PFU)

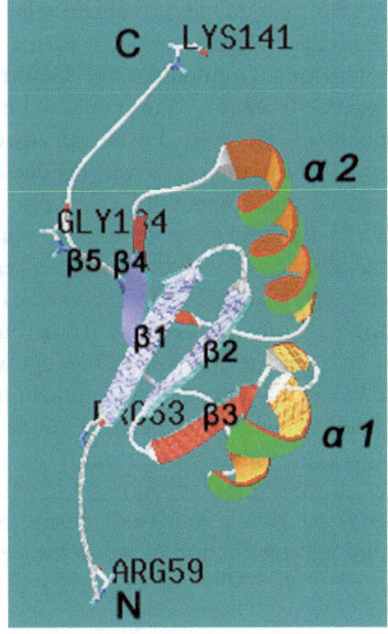

respectively to map the regions on ExbD involved in TonB interactions (Garcia-Herrero et al. 2007). The NMR chemical shift perturbation strategy to study binding interactions showed that the addition of Glu-Pro peptide to monomeric ExbD caused several peaks to shift at pH 3.0 indicating multiple interactions with ExbD, while the addition of Ly-Pro peptide did not cause any change. Further analysis showed that these interactions mapped on multiple sites on ExbD and were weaker in nature. However, the same experiments performed with dimeric ExbD at pH 7.0 did not show any interactions, rejecting the possibility of inter-actions in vivo according to Garcia-Herrero et al. (Garcia-Herrero et al. 2007). However, Ollis and Postle recently identified regions of ExbD involved in the formation of homodiemer of ExbD and heterodiemers of ExbD-TonB based on cysteine scanning involving residues 92–121 of ExbD (Ollis and Postle 2011). These regions were mapped in the same regions identified by Vogel and group, while few regions were located on the opposite side (Ollis and Postle 2011, Garcia-Herrero et al. 2007). Some of the residues involved in the formation of a homodiemer of ExbD were also involved in ExbD-TonB complexs formation via a disulfide bond formation between the cysteine residue of ExbD and A150C of TonB (Ollis and Postle 2011). The TonB residue A150, however, did not fall in the regions studied by Garcia-Herrero et al. In a recent study, FhuD was shown to form a 1:1 complex with CTD of TonB (Carter et al. 2006). The binding site for TonB on FhuD protein, mapped based on the random phage peptide display method, was located near the ferrichrome binding site within the cleft formed by the globular C-terminal and N-terminal domains of FhuD and not on the C-terminal lobe which shares structural folding homology with ExbD (Carter et al. 2006). As FhuD and ExbD proteins have two different and distinct roles to play, it is highly unlikely for them to share the same binding sites on TonB if they are interacting simultaneously. Unfortunately, the sequence of events during the process of energy transduction and transport involving these proteins are largely unknown. Recently, Postle and coworkers showed that the transmembrane domain, specifically, the conserved residue D25 and the N-terminal flexible region of the periplasmic domain of ExbD (residues 42–61) are necessary for bringing about the PMF-dependent conformational changes on TonB (Ollis and Postle 2012a, b). In the absence of the periplasmic N-terminal flexible region or in the D25 N mutant, the periplasmic C-terminal residues (64–141) of ExbD, which form a well-defined folded structure, binds to TonB but blocks the PMF-dependent energized inter-actions of the TonB-ExbD complex (Ollis and Postle 2012a, b). The studies also indicate that the physical interaction between the periplasmic domain of ExbD and TonB is energy independent and is a prior requirement for the subsequent step of energy transduction to occur (Ollis and Postle 2012a, b). Previous studies by Braun et al. (1996) and recently by Ollis and Postle showed that the residues L123 and H20 on ExbD and TonB respectively are important for the PMF-independent physical interaction between the periplasmic domains of TonB and ExbD (Braun et al. 1996; Ollis and Postle 2012a, b).

1.4.2 ExbB Structure

Even though ExbB is the major protein in the TonB-ExbB-ExbD complex, it has not been investigated thoroughly. Braun and coworkers recently carried out structural analysis of ExbB and ExbB-ExbD complex using chromatographic and mass-spectroscopic (laser-induced liquid bead ion desorption mass spectrometry) analysis of detergent (undecyl maltoside) purified protein (Pramanik et al. 2011). Their analysis indicated that ExbB mainly exists as a hexameric form with minor amounts of trimeric, dimeric, and monomeric forms, while the prevalent ExbB-ExbD stoichiometry was 6:1 along with traces of 3:1, 2:1 and 1:1. The hexamer of ExbB and ExbB-ExbD complex in a 6:1 ratio was found to the most stable forms. ExbB and ExbD proteins play a role in energizing the TonB by coupling the PMF of the cytoplasmic membrane. They are homologs of the MotA and MotB proteins found in the bacterial flagellar motor (Braun and Braun 2002b) and also the TolQ and TolR proteins involved in energizing group A colicin uptake (Germon et al. 2001). There is a wealth of structural and biochemical information available on TonB, ExbB, ExbD, and about their physical interaction; however, this has not helped to resolve the pathway of energy transduction. If the crystal structure of the whole TonB-ExbB-ExbD complex can be determined, it would certainly help to get more structural information about their interactions. This may not, however, solve the mystery of the process of energy transduction but it may provide ground to further design experiments to understand the mechanism of energy transduction. There is number of limitations for achieving this goal mainly due to a difficulty in obtaining the full length TonB and ExbD protein in a stable form. Although a recent report by Postle's group validated a use of a spheroplast as a system to study the interactions between TonB and ExbD in vivo. This may help to carry out meaningful experiments in the future to resolve the mystery of the energy transfer pathway (Ollis and Postle 2012a, b).

1.4.3 Role of TonB-ExbB-ExbD Complex in Energy Transduction

Outer membranes in Gram-negative bacteria have a number of active transport units transporting ferric-siderophores, antibiotics, colicins, phages, etc., but are devoid of any energy source to support active transport. There are a number of structural and biochemical reports which provide evidence for the involvement of TonB-ExbB-ExbD protein complex located in the inner membrane in energy transduction. None of these reports, however, reveal the exact pathway of energy transfer between either the components of the TonB-ExbB-ExbD complex or between the complex and TBDTs. The energy required for the opening of the channel in TBDTs is provided by the proton motive force created by the electrochemical potential of the inner membrane. This is proven by the fact that protonophores, which sequester protons, strongly inhibit TonB-dependent transports. According to current understanding, the binding of ferric-siderophore to the TBDTs results in the conformational changes initiating the energy-dependent step by extending and making the TonB box available for interaction with the C-terminal TonB. A number of biochemical, site-directed mutagenesis and formaldehyde cross-linking experiments have proven that TonB interacts physically with the N-terminal domain and that this interaction is necessary for the energy driven second step of the ferric-siderophore transport. The nature of the physical interaction was subsequently revealed with help of the crystal structures of FhuA-C-terminal TonB and BtuB- C-terminal TonB complexes as described in the previous sections (Pawelek et al. 2006; Shultis et al. 2006). Both structures showed that the C-terminus is swapped for TonB box of the OM receptor protein. As previously mentioned, the structures neither reveal nor give any clue about the mechanism of the energy transduction. The energy transduction involves interaction of TonB with two other inner membrane proteins, ExbB and ExbD. The TonB-ExbB-ExbD complex is also involved in energizing the transport of heme, antibiotics, sugars, and nickel (Braun and Braun 2002a; Ferguson and Deisenhofer 2004; Schauer et al. 2007). ExbD and TonB share a common structural feature in that both the proteins have single transmembrane domains and the majority of their C-terminal domain is located in the periplasm. ExbB, on the other hand, traverses the membrane thrice and has a large portion of its C-terminal domain located in the cytoplasm. It has been reported that TonB cycles through a set of conformations that differ in potential energy, with a transition from a PMF driven higher energy state to a lower energy state, releasing potential energy to an outer membrane receptor (Larson et al. 1999). Postle and coworkers have recently published their work on ExbB and ExbD revealing the roles of specific regions of ExbB and ExbD based on elegant experiments using 10 residue deletions, site-directed mutagenesis and also spheroplasts (Ollis and Postle 2012a, b; Ollis et al. 2012). They have also proposed a model for the sequence of events leading to TonB energization. There have been many hypotheses about the way TonB energizes the OM receptors, including the hypothesis that it acts as a homodimer or the hypothesis that it shuttles between the membranes to deliver energy. These hypotheses were recently disproved (Ködding et al. 2004; Gresock et al. 2011; Postle et al. 2010). According to the data published by the Postle's group and other reports, TonB undergoes three different conformations in vivo during the process of energy transfer, named stages I to III. Stage I and II are PMF-independent while stage III is PMF dependent (Ghosh and Postle 2005, Gresock et al. 2011; Ollis and Postle 2012a, b). Stage I consists of the recognition of periplasmic TonB by the C-terminal domain of ExbD. This interaction is inhibited by mutations in H20 and L132 of TMD of TonB and ExbD, respectively (Braun et al. 1996; Ollis and Postle 2012a, b). H20A mutation might not allow the conformational change in TonB necessary for it to be accessed by the CTD of ExbD. On the other hand, L132 might be directly involved in the physical interaction with TonB and its mutation to Q might prevent the interactions (Ollis and Postle 2012a, b). The TonB at this stage is still proteinase K sensitive (indicating no major conformational change) (Ollis and Postle 2012a, b). In stage II, TonB physically interacts closely with the C-terminal domain (62–141 residues) of ExbD, proven based on disulfide bridge formation, deletions, and formaldehyde linkage experiments. TonB also becomes proteinase K resistant during this stage (Ollis and Postle 2012a, b). This stage is PMF-independent proven by the fact that it is not inhibited by protonophores like CCCP. Stage II interactions are not affected in the D25N mutant which shows the same phenotype as in the presence of CCCP. Stage III is defined by the PMF-dependent conformational changes in TonB where protein again becomes proteinase K sensitive and is able to form PMF-dependent formaldehyde linkages. As expected, this stage is strongly inhibited in the presence of CCCP as well as in D25 N mutant (Ollis and Postle 2012a, b). This model is very convincing, however, there are a number of questions unanswered: (1) we now know that D25 in ExbB TMD is involved in generating PMF, but what is the complete proton transfer pathway? (2) What are the sequence of events and the exact nature of the physical interactions between TonB and ExbD that energizes TonB? (3) What is the exact role(s) of ExbB? and (4) How exactly is the energy transduced to the OM receptor? It was reported that mutations in the cytoplasmic domain of ExbB interferes with the physical interactions between the periplasmic domains of TonB and ExbD, however, this mechanism is still unknown. The mechanism by which mutations/deletions in a short N-terminal cytoplasmic domain of ExbD prevents interactions of TonB-ExbD is still unknown. Do these regions on ExbB and ExbD interact with each other? Or do they play any role in generating PMF? Are they simply important for protein assembly? Answers to these questions are vital to the understanding of the mechanism.

1.5 Structures of Components Involved in Genetic Regulation

The genes involved in expressing siderophore biosynthetic and the transport components are under strict regulation as maintaining iron concentration is critical for the survival of cells. Iron at high concentration is toxic to the cells. This is due to its ability to catalyze a Fenton-type reaction producing highly reactive oxygen species (e.g. hydroxyl ions, Todd et al. 2002). Therefore, the transport of iron is tightly regulated at the genetic level (Crosa 1989, 1997; Frost and Rosenberg 1973). In *E. coli* and many other bacteria the 'Fur' (ferric uptake regulator) protein is a global regulator which is an iron-responsive DNA-binding repressor protein that requires Fe^{+2} as a co-repressor (Bagg and Neilands 1987). It binds to a 19 bp long palindromic consensus sequence (Fur box) near the promoters of iron-regulated genes. In *E. coli* and *Pseudomonas* the Fur regulon regulates at least 90 genes including the genes involved in siderophore-mediated iron transport (Wexler et al. 2003). Besides regulating iron uptake mechanisms, Fur has been found to be involved in regulation of an alternative sigma factor, as well as regulation of activator genes, stress response genes, energy metabolism genes, and also virulence associated genes (Chao et al. 2004). The first crystal structure of Fur from *P.aeruginosa* was solved by Pohl et al. in 2003 (Pohl et al. 2003). Recently in 2009, a structure of Fur protein from pathogenic bacterium *Vibrio cholerae* was solved by Sheikh and Taylor (Sheikh and Taylor 2009). The crystal structure of Fur proteins revealed that it is a dimeric metal binding protein which has a classic winged helix motif at the N-terminal for DNA binding (Pohl et al. 2003; Sheikh and Taylor 2009). The structures also showed the presence of two Zn^{+2} atoms per monomer, one of which plays a regulatory role in activation of Fur protein while the other has a structural role (Pohl et al. 2003; Sheikh and Taylor 2009). In addition to the Fur mediated control, in the case of a few ferric-siderophore transport systems, the gene expression is also positively controlled by the extracytoplasmic factors (ECF), e.g. ferric-dicitrate transport via FecA in *E. coli* (Garcia-Herrero and Vogel 2005). The ferric-dicitrate transporter FecA along with two other proteins FecR and FecI, located in the inner membrane and cytoplasm respectively, are involved in a signaling mechanism which allows the system to be controlled positively in the presence of ferric citrate. The positively regulated transporters such as FecA and FpvA transporting ferric dicitrate and ferric pyoverdine respectively have extended N-terminal domain (NTD) into the periplasm. The NTD is involved in ECF signaling (Garcia-Herrero and Vogel 2005). For example, binding of ferric dicitrate to FecA initiates conformational changes into the transporter leading to the interaction of NTD with FecR, a regulator located in the inner membrane. This in turn activates a cytoplasmic sigma factor FecI, which initiates the transcription of FecABCDE genes (Sen et al. 2008). The outer membrane receptors involved in ECF signaling have about 80 extra residues at their N-terminal domain, which interacts with the periplasmic domain of the inner membrane protein during signaling. These extended residues were not detected in the electron density maps of the crystal structures of either FecA or FpvA due to a high degree of flexibility and disorder in their configuration (Ferguson et al. 2002, 2006; Garcia-Herrero and Vogel 2005). Recently, Garcia-Herrero and Vogel and Ferguson et al. have solved the solution-structure of the extended NTD of FecA containing residues 1–79 along with the residues 80–96 of the plug domain (residues 1–80, Ferguson et al. 2006) by multidimensional nuclear magnetic resonance (NMR) spectroscopy. The structure of residues 1–74 showed that it was well ordered, with the core region of the extended domain containing two alpha-helices flanked by two beta-sheets on one side and three on the other. The first beta-sheet is formed by two beta strands linked with a small alpha-helical turn (Garcia-Herrero and Vogel 2005, Ferguson et al. 2006). The region that connected the extended domain to the plug domain and contains TonB box (residues 80–84) was found to be unstructured and flexible by Garcia-Herrero and Vogel. This had already been observed in the case of the liganded crystal structures of FecA (Ferguson et al. 2002).

1.6 Mechanism of Transport via Outer Membrane

In spite of having considerable information based on structural, biochemical, and genetic analyses on the ferric-siderophore transport system, the mechanism of transport via TBDTs remains elusive. I have discussed the structural and to some extent biochemical information in the previous sections, but in the last decade there have been some important observations made on the biophysical properties of the ferric-siderophore receptors that have not been mentioned (Lo Conte et al. 1999; Tsai et al. 1997, Chimento et al. 2005). The biophysical analysis combined with the information above may help us to understand the sequence of events during transport. For example, according to the studies by Chimento et al., the nature of the protein–protein interaction between the barrel and the plug domain of TBDTs is like transient protein complexes that undergo conformational changes or domain movement during their function (Chimento et al. 2005). This is based on their observation that the interfaces are very large and contain large numbers of water molecules. However, the number of water molecules that form hydrogen bonds with both the barrel and a plug domain, called bridging water molecules, is only one-third of the total. The typical value of hydrogen bonds per water molecule is 3.8 for interfacial water but in this case it is 3.1. The barrel-plug domain interface contains nearly 40 % polar residues, which is typical of the transient protein complexes (Lo Conte et al. 1999, Tsai et al. 1997). These observations suggest that the plug domain can undergo conformational changes with a minimal energy cost. This supports the hypothesis that the plug domain does not need to come out, but can undergo conformational changes to facilitate the delivery of the ferric-siderophore complex into the periplasm. Studies by Gumbart et al. (Gumbart et al. 2007) using molecular dynamics simulations on the BtuB-TonB interactions found that the attachment of TonB to the BtuB is strong enough to pull and unfold the plug domain enough to allow the translocation of the substrate. It would need to be 10 times stronger to pull out the whole plug. The results are in line with the data obtained by Chimento et al. (Chimento et al. 2005). Double cysteine mutants created to hold the plug inside the barrel, showed close to normal binding and transport, supporting these observations (Eisenhauer et al. 2005; Chakraborty et al. 2007). In order to understand the mechanism of a possible conformational change in the plug domain, Chakraborty et al. identified several conserved residues among the TBDTs. A cluster of 10 residues, which we called a 'lock region', consisted of a few charged residues and glycines from the plug and barrel domains. The charged residues, like glutamates and arginines, formed salt bridges while the conserved glycines located on the bending of the helices could serve as a hinge to facilitate conformational changes. Mutation of several of these residues to alanine in both FepA and FhuA severely affected transport but not binding (Chakraborty et al. 2003, 2007). The double cysteine mutant where the central β-strands 4 and 5 of FhuA were tied preventing their possible movement during the conformational change showed completely diminished transport (Chakraborty et al. 2007). The reports strongly support the idea that the whole plug does not need to come out and that the dislodgment of the plug domain is energetically not a favorable proposition. However, there are a number of questions that are unanswered regarding the sequence of events that possibly bring about the conformational change in the plug domain. First, what causes the conformational change following the binding of ligand to the receptor leading to the interaction of TonB box with TonB, that forms a pore large enough to allow the passage of the ligand? Second, are there other allosteric signaling mechanisms besides the interaction of TonB with TonB box that induces the conformational change in the plug domain? And third, what is the sequence of events that restores the intact plug to its original position? It was reported that the TMD of TonB which contains the conserved residue H20 is not involved in proton transfer. To date D25 of ExbD is the only residue proven to be involved in the proton transfer (Ollis and Postle 2012a, b). It has also been proven that ExbD physically interacts with TonB; therefore, it can be assumed that ExbD probably transfers protons via its interaction with TonB, which eventually pumps protons into the 'lock region' thereby disrupting the salt bridges. Subsequently, the conserved glycines facilitate the movements of the β-strands of the plug domain to form a pore. The biochemical data on binding and transport using the mutants involving the 'lock region' support this hypothesis (Chakraborty et al. 2003, 2007). A diminished transport due to tying of the central β-strands of the plug domain with the help of a double cysteine mutant (N104C/L149C) of FhuA provides further evidence for a conformational change (Chakraborty et al. 2007). We further analyzed the double cysteine mutant to verify whether the mutation had effects on its interaction with TonB leading to a severely diminished transport. The double cysteine mutant showed normal ferrichrome-dependent interaction with C-terminal TonB (unpublished data). This means that the diminished transport in the double cysteine mutant was not due to the lack of its ability to interact with TonB but it was indeed due to its inability to bring about a conformational change in the 'lock region'. Once the ligand moves across the receptor, it is received by the periplasmic binding protein where TonB might act as a scaffold facilitating a transfer of the ligand to PBP. The PBP then delivers the ligand to the ABC type inner membrane transporter to be transported into the cytoplasm using ATP as an energy source. Recently, *Carter* et al. reported that TonB acts as a scaffold, directing a periplasmic ferrichrome binding protein, FhuD, to regions within the periplasm where it is poised to accept and deliver siderophore (Carter et al. 2006). FhuD structure was solved in complex with several different ligands (Clarke et al. 2000, 2002). FhuD facilitates the transfer of ligand from FhuA to the inner membrane ATP-binding cassette (ABC) type transporter. ABC type transporters are ATP-dependent inner membrane permeases which transport the iron bound ligands into the cytoplasm (Locher and Borths 2004). In the cytoplasm, iron is released from the ligand either via enzymatic reduction of ferric ion to ferrous ion or via degradation of the siderophore molecule.

1.7 Conclusions

Despite substantial information gathered over the past decade on the structural components of the ferric-siderophore transport systems, the mystery regarding the exact mechanism of ligand passage through the TBDTs and the steps involved in energy transduction has not yet been solved. The major problem is posed by involvement of the large numbers of components spanning the outer membrane to the cytoplasm in the transport process. The limitation of structural data is that it represents a solid state, and therefore only captures a single conformation. The biochemical and genetic analyses involving either incomplete polypeptides or mutated ones, on the other hand, can generate misleading data due to unpredictable structural changes caused by the mutations or misfolding in incomplete poly-peptides. The recently reported use of spheroplast as an in vivo system may help to design meaningful experiments in the future. Neither structural, biochemical, nor genetic experiments alone will solve the mystery surrounding membrane transport. It will require a combination of experiments from many perspectives.

References

Abergel RJ, Clifton MC, Pizarro JC et al (2008) The siderocalin/enterobactin interaction: a link between mammalian immunity and bacterial iron transport. J Am Chem Soc 130:11524–11534

Abergel RJ, Moore EG, Strong RK et al (2006) Microbial evasion of the immune system: structural modifications of enterobactin impair siderocalin recognition. J Am Chem Soc 128:10998–10999

Bagg A, Neilands JB (1987) Molecular mechanism of siderophore-mediated iron assimilation. Microbiol Rev 51:509–518

Barnard TJ, Watson ME Jr, McIntosh MA (2001) Mutations in E. coli receptor FepA reveal residues involved in ligand binding and transport. Mol Microbiol 41:527–536

Barnes CL, Eng-Wilmot DL, van der Helm D (1984) Ferricrocin ($C_{29}H_{44}FeN_9O_{13}.7H_2O$), an iron(III)-binding peptide from Aspergillus versicolor. Acta Cryst C40:922–926

Barnes CL, Hossain MB, Jalal MAF et al (1985) Ferrichrome conformations: ferrirubin, two crystal forms: $C_{41}H_{64}FeN_9O_{17}.10.5H_2O$ (I) and $C_{41}H_{64}FeN_9O_{17}.CH_3CN.H_2O$ (II). Acta Cryst C41:341–347

Borths EL, Locher KP, Lee AT et al (2002) The structure of E. coli BtuF and binding to its cognate ATP binding cassette transporter. Proc Natl Acad Sci USA 99:16642–16647

Bradbeer C (1993) The proton motive force drives the outer membrane transport of cobalamin in Escherichia coli. J Bacteriol 175:3146–3150

Braun V (1995) Energy-coupled transport and signal transduction through the Gram-negative outer membrane via TonB-ExbB-ExbD-dependent receptor proteins. FEMS Microbiol Rev 16:295–307

Braun V (1997) Surface signaling: novel transcription initiation mechanism starting from the cell surface. Arch Microbiol 167:325–331

Braun V, Braun M (2002a) Active transport of iron and siderophore antibiotics. Curr Opinion Microbiol 5:194–200

Braun V, Braun M (2002b) Iron transport and signaling in E. coli. FEBS Lett 529:78–85

Braun V, Gaisser S, Herrmann C et al (1996) Energy-coupled transport across the outer membrane of *E. coli*: ExbB binds ExbD and TonB in vitro, and leucine 132 the periplasmic region and aspartate 25 in the transmembrane region are important for ExbD activity. J Bacteriol 178:2836–2845

Braun V, Hantke K, Koster W (1998) Bacterial iron transport: mechanism, genetics and regulation. In: Sigel A, Sigel H (eds) Metal ions in biological systems. Marcel Dekker, Inc., New York, pp 67–145

Braun V, Mahren S, Ogierman M (2003) Regulation of FecI-type ECF sigma factor by transmembrane signaling. Curr Opin Microbiol 6:173–180

Buchanan SK, Smith BS, Venkatramani L et al (1999) Crystal structure of the outer membrane active transporter FepA from *Escherichia coli*. Nature (Struct Biol) 6:56–63

Carter DM, Miousse IR, Gagnon JN et al (2006) Interactions between TonB from *E. coli* and the periplasmic protein FhuD. J Biol Chem 281:35413–35424

Chakraborty R, Lemke EA, Cao Z et al (2003) Identification and mutational studies of conserved amino acids in the outer membrane receptor protein, FepA, which affect transport but not binding of ferric-enterobactin in *E. coli*. Biometals 16:507–518

Chakraborty R, Storey E, van der Helm D (2007) Molecular mechanism of ferricsiderophore passage through the outer membrane receptor proteins of *E. coli*. Biometals 20:263–274

Chang C, Mooser A, Pluckthun A et al (2001) Crystal structure of the dimeric C-terminal domain of TonB reveals a novel fold. J Biol Chem 276:27535–27540

Chao TC, Becker A, Buhrmester J et al (2004) The *Sinorhizobium meliloti* fur gene regulates, with dependence on Mn(II), transcription of the sitABCD operon, encoding a metal-type transporter. J Bacteriol 186:3609–3620

Chimento DP, Kadner RJ, Wiener MC (2005) Comparative structural analysis of TonB-dependent outer membrane transporters: implications for the transport cycle. Protein Struct Funct Bioinformatics 59:240–251

Chimento DP, Mohanti AK, Kadner RJ et al (2003) Substrate-induced transmembrane signaling in the cobalamin transporter BtuB. Nat Struct Biol 10:394–401

Chu BC, Garcia-Herrero A, Johanson TH et al (2010) Siderophore uptake in bacteria and the battle for iron with the host; a birds's eye view. Biometals 23:597–599

Clarke TE, Braun V, Winkelmann G et al (2002) X-ray crystallographic structures of the *Escherichia coli* periplasmic protein FhuD bound to hydroxamate-type siderophores and the antibiotic albomycin. J Biol Chem 277:13966–13972

Clarke TE, Ku S, Dougan DR et al (2000) The structure of the ferric siderophore binding protein FhuD complexed with gallichrome. Nat Struct Biol 7:287–291

Cobessi D, Celia H, Pattus F (2005a) Structure of ferric-pyochelin and its membrane receptor FptA from *P. aeruginosa*. J Mol Biol 352:893–904

Cobessi D, Herve C, Folschweiller N et al (2005b) The Crystal structure of the pyoverdine outer membrane receptor FpvA from *P. aeruginosa* at 3.6 Å resolution. J Mol Biol 347:121–134

Crosa JH (1989) Genetics and molecular biology of siderophore-mediated iron transport in bacteria. Microbiol Rev 53:517–530

Crosa JH (1997) Signal transduction and transcriptional and postranscriptional control of iron-regulated genes in bacteria. Microbiol Mol Biol Rev 61:319–336

de Lorenzo V, Herrero M, Giovannini F et al (1988) Fur (ferric uptake regulation) protein and CAP (catabolite-activtor protein) modulate transcription of fur gene in *Escherichia coli*. Eur J Biochem 173:537–546

Eisenhauer HA, Shames S, Pawelek PD et al (2005) Siderophore transport through *E. coli* outer membrane receptor FhuA with disulfide-tethered cork and barrel domains. J Biol Chem 280:30574–30580

Ferguson A, Chakraborty R, Barbara S et al (2002) Structural basis of gating by the outer membrane transporter FecA. Science 295:1715–1719

Ferguson AD, Deisenhofer J (2004) Metal import through microbial membranes. Cell 116:15–24

Ferguson AD, Hofmann E, Coulton JW et al (1998) Siderophore-mediated iron transport: crystal structure of FhuA with bound lipopolysaccharide. Science 282:2215–2220

Ferguson AD, Amezcua CA, Halabi NM et al (2006) Signal transduction pathway of TonB-dependent transporters. Proc Natl Acad Sci USA 104:513–518

Francis JJ, Macturk HM, Madinaveitia J et al (1949) Isolation from acid-fast bacteria of a growth-factor for *Mycobacterium johnei* and of a precursor of phthiocol. Nature 163:365–366

Frost GE, Rosenberg H (1973) The inducible citrate-dependent iron transport system in *Escherichia coli* K12. Biochem Biophys Acta 330:90–101

Garcia-Herrero A, Peacock RS, Howard SP et al (2007) The solution structure of the periplasmic domain of the TonB system ExbD protein reveals an unexpected structural homology with siderophore-binding proteins. Mol Microbiol 66:872–889

Garcia-Herrero A, Vogel HJ (2005) Nuclear magnetic resonance solution structure of the periplasmic signaling domain of the TonB-dependent outer membrane transporter FecA from *E. coli*. Mol Microbiol 58:1226–1237

Germon P, Ray MC, Vianney A et al (2001) Energy-dependent conformational change in the TolA protein of *E. coli* involves its N-terminal domain, TolQ and TolR. J Bacteriol 183: 4110–4114

Ghosh J, Postle K (2005) Disulphide trapping of an in vivo energy-dependent conformation of *Escherichia coli* TonB protein. Mol Microbiol 55:276–288

Gresock MG, Savenkova MI, Larsen RA et al (2011) Death of the TonB shuttle hypothesis. . doi:10.3389/fmicb.2011.00206

Gudmundsdottir A, Bell PE, Lundrigan MD et al (1989) Point mutations in a conserved region (TonB box) of *Escherichia coli* outer membrane protein BtuB affect vitamin B12 transport. J Bacteriol 171:6526–6533

Gumbart J, Weiner MC, Tajkhorshid E (2007) Mechanics of force propagation in TonB dependent outer membrane transport. Biophysical J 93:496–504

Hantke K (1984) Cloning of the repressor protein gene of iron-regulated systems in *Escherichia coli* K12. Mol Gen Genet 197:337–341

Higgs PI, Larsen RA, Postle K (2002) Quantification of known components of the *Escherichia coli* TonB energy transduction system: TonB, ExbB, ExbD and FepA. Mol Microbiol 44: 271–281

Hough E, Rogers D (1974) Crystal structure of ferrimycobactin P, a growth factor for the mycobacteria. Biochem Biophys Res Commun 57:73–77

Jalal MA, Mocharla R, Barnes CL et al (1984a) Extracellular siderophores from *Aspergillus ochraceous*. J Bacteriol 158:683–688

Jalal MA, Mocharla R, van der Helm D (1984b) Separation of ferrichromes and other hydroxamate siderophores of fungal origin by reversed-phase chromatography. J Chromat 301:247–252

Jalal MAF, van der Helm D (1989) Purification and crystallization of ferric enterobactin receptor protein, FepA, from the outer membranes of *E. coli* UT5600/pbb2. FEBS Lett 243:366–370

Jana B, Manning M, Postle K (2011) Mutations in the ExbB cytoplasmic carboxy terminus prevent energy-dependent interaction between the TonB and ExbD periplasmic domains. J Bacteriol 193:5649–5657

Killmann H, Videnov G, Jung G et al (1995) Identification of receptor binding sites by competitive peptide Mapping: phages T1, T5, and 80 and colicin M bind to the gating loop of FhuA. J Bacteriol 177:694–698

Ködding J, Killig F, Howard SP et al (2004) Dimerization of TonB is not essential for its binding to the outer membrane siderophore receptor FhuA of *E. coli*. J Biol Chem 279:9978–9986

Ködding J, Killig F, Polzer P et al (2005) Crystal structure of a 92-residue C-terminal fragment of TonB from *E. coli* reveals significant conformational changes compared to structures of smaller TonB fragments. J Biol Chem 280:3022–3088

Krewulak KD, Vogel HJ (2008) Structural biology of bacterial iron uptake. Bioch Biophys Acta 1778:1781–1804

Larson RA, Letain TE, Postle K (1999) Protonmotive force, ExbB and ligand-bound FepA drive conformational changes in TonB. Mol Microbiol 31:1809–1824

Lo Conte L, Chothia C, Janin J (1999) The atomic structure of protein–protein recognition sites. J Mol Biol 285:2177–2198

Locher KP, Borths E (2004) ABC transporter architecture and mechanism: implications from the crystal structures of BtuCD and BtuF. FEBS Lett 564:264–268

Locher KP, Rees B, Koebnik R et al (1998) Transmembrane signaling across the ligand-gated FhuA receptor: crystal structures of free and ferrichrome-bound states reveal allosteric changes. Cell 95:771–778

Lundrigan MD, Kadner RJ (1986) Nucleotide sequence of the gene for the ferrienterochelin receptor FepA in *E. coli*. J Biol Chem 261:10797–10801

Neilands JB (1981) Microbial iron compounds. Ann Rev Biochem 50:715–731

Neilands JB (1952) A crystalline organo-iron pigment from the smut fungus *Ustilago sphaerogena*. J Am Chem Soc 74:4846–4847

Newton SMC, Igo JD, Scott DC et al (1999) Effect of loop deletion on the binding and transport of ferric enterobactin by FepA. Mol Microbiol 32:1153–1165

Noinaj N, Guillier M, Barnard TJ et al (2010) TonB dependent transporters: regulation, structure, and function. Annu Rev Microbiol 64:43–60

Ollis AA, Kumar A, Postle K (2012) The ExbD periplasmic domain contains distinct functional regions for two stages in TonB energization. J Bacteriol 194:3069–3077

Ollis AA, Postle K (2011) The same periplasmic ExbD residues mediate in vivo interactions between ExbD homodimers and ExbD-TonB heterodimers. J Bacteriol 193:6852–6863

Ollis AA, Postle K (2012a) Identification of functionally important TonB-ExbD periplasmic domain interactions in vivo. J Bacteriol 194:3078–3087

Ollis AA, Postle K (2012b) ExbD mutants define initial stages in TonB energization. J Mol Biol 415:237–247

Pawelek PD, Croteau N, Ng-Thow-Hing C et al (2006) Structure of TonB in complex with FhuA, *E. coli* outer membrane receptor. Science 312:1399–1402

Peacock RS, Weljie AM, Howard SP et al (2005) The solution structure of the C-terminal domain of TonB and interactions studies with TonB box peptides. J Mol Biol 345:1185–1197

Pohl E, Haller J, Mijovilovich A et al (2003) Architecture of a protein central to iron homeostasis: crystal structure and spectroscopic analysis of the ferric uptake regulator. Mol Microbiol 47:903–915

Postle K (1993) TonB protein and energy transduction between membranes. J Bioenerg Biomembr 25:591–601

Postle K, Kastead K A, Gresock MG et al. (2010) The TonB dimeric crystal structures do not exist in vivo. mBio 1:e00307–003010. doi:10.1128/mBio.00307-10

Pramanik A, Hauf W, Hoffmann J et al (2011) Oligomeric structure of ExbB and ExbB-ExbD isolated from *E. coli* as revealed by LILBID mass spectrometry. Biochemistry 50:8950–8956

Pramanik A, Zhang F, Schwarz H, Schreiber F et al (2010) ExbB protein in the cytoplasmic membrane of *E. coli* forms a stable oligomer. Biochemistry 49:8721–8728

Raymond K, Dertz RE, Kim SS (2003) Enterobactin: an archetype for microbial iron transport. Proc Natl Acad Sci 100:3584–3588

Rousseau F, Schymkowitz JW, Itzhaki LS (2003) The unfolding story of three-dimensional domain swapping. Structure (Camb.) 11:243–251

Sauer M, Hantke K, Braun V (1990) Sequence of the fhuE outer-membrane receptor gene of Escherichia coli K12 and properties of mutants. Mol Microbiol 4:427–437

Sauter A, Howard SP, Braun V (2003) In vivo evidence for TonB dimerization. J Bacteriol 185:5747–5754

Schauer K, Gouget B, Carriere M et al (2007) Novel nickel transport mechanism across the bacterial outer membrane energized by the TonB/ExbB/ExbD machinery. Mol Microbiol 63:1054–1068

Sen TZ, Kloster M, Jemigan RL et al (2008) Predicting the complex structure and functional motions of the outer membrane transporter and signal transducer FecA. Biophysical J 94:2482–2491

Sheikh A, Taylor GL (2009) Crystal structure of the *Vibrio cholerae* ferric uptake regulator (Fur) reveals insights into metal co-ordination. Mol Microbiol 72:1208–1220

Shultis DD, Purdy MD, Banchs CN et al (2006) Outer membrane active transport: Structure of the BtuB: TonB complex. Science 312:1396–1399

Skare JT, Ahmer BMM, Seachord CL et al (1993) Energy transduction between membranes. J Biol Chem 268:16302–16308

Teintze M, Hossain MB, Barnes CL et al (1981) Structure of ferric pseudobactin, a siderophore from a plant growth promoting *Pseudomonas*. Biochemistry 20:6446–6457

Todd JD, Wexler M, Sawers G et al (2002) RirA, an iron-responsive regulator in the symbiotic bacterium *Rhizobium leguminosarum*. Microbiology 148:4059–4071

Tsai CJ, Lin SL, Wolfson HJ et al (1997) Studies of protein–protein interfaces: a statistical analysis of the hydrophobic effect. Protein Sci 6:53–64

Twort FW, Ingram GLY (1912) A method for isolating and cultivating *Mycobacterium enteritidis chronicae pseudotuberculosis bovis, Johne*, and some experiments on the preparation of a diagnostic vaccine for pseudotuberculosis enteritis of bovines. Proc R Soc Lond Ser B 84:517–542

Vakharia-Rao H, Kastead KA, Savenkova MI et al (2007) Deletion and substitution analysis of *E. coli* TonB Q160 region. J Bacteriol 189:4662–4670

van der Helm D (2004) Structure of outer membrane receptor proteins pp 51–65. In: Crosa JH, Mey AR, Payne SM (eds) Iron transport in bacteria. ASM press, Washington

van der Helm D, Baker JR, Eng-Wilmot DL et al (1980) Crystal structure of ferrichrome and a comparison with the structure of ferrichrome A. J Am Chem Soc 102:4224–4231

van der Helm D, Baker JR, Loghry RA et al (1981) Structures of alumichrome A and ferrichrome A at low temperature. Acta Cryst B37:323–330

van der Helm D, Chakraborty R (2002) Structures of siderophore receptors. In: Winkelmann G (ed) Microbial transport systems. Wiley, Weinheim, pp 261–288

van der Helm D, Chakraborty R, Ferguson AD et al (2002) Bipartite gating in the outer membrane protein FecA. Biochem Soc Trans 30:708–710

van der Helm D, Poling M (1976) The crystal structure of ferrioxamine E. J Am Chem Soc 98:82–86

Wexler M, Todd JD, Kolade O et al (2003) Fur is not the global regulator of iron uptake genes in *Rhizobium leguminosarum*. Microbiology 149:1357–1365

Chapter 2
The Tricky Ways Bacteria Cope with Iron Limitation

Volkmar Braun and Klaus Hantke

Abstract Iron is an essential element for many key redox systems. It is difficult to acquire for cells under oxic conditions, since Fe^{3+} forms insoluble hydroxides. In the human host, iron is tightly bound to proteins. Bacteria invented iron transport systems which solubilize external Fe^{3+} by secreted low-molecular weight compounds, designated siderophores, or directly from the human proteins. Gram-negative bacteria contain an intricate energy-coupled iron transport mechanism across the outer membrane which lacks an energy source. The electrochemical potential of the cytoplasmic membrane delivers the energy. Transport across the cytoplasmic membrane is most frequently achieved by ABC transporters in Gram-positive and Gram-negative bacteria. Under anaerobic conditions, iron is the soluble Fe^{2+} form and transported different to Fe^{3+}. Iron transport and intracellular iron concentrations are controlled by transcription regulation of iron transport genes. Transcription is turned on under iron-limiting growth conditions which usually exist in natural environments.

Keywords Bacterial iron transport · Siderophores · Human iron binding proteins

V. Braun (✉)
Max-Planck-Institute for Developmental Biology,
Spemannstrasse 35, Tübingen 72076, Germany
e-mail: volkmar.braun@tuebingen.mpg.de

K. Hantke
IMIT, Universität Tübingen, Auf der Morgenstelle 28,
Tübingen 72076, Germany
e-mail: hantke@uni-tuebingen.de

R. Chakraborty et al. (eds.), *Iron Uptake in Bacteria with Emphasis on E. coli and Pseudomonas*, SpringerBriefs in Biometals, DOI: 10.1007/978-94-007-6088-2_2, © The Author(s) 2013

2.1 Introduction

For almost all bacteria, iron is an essential nutrient since it is contained in the redox centers of many enzymes of the respiratory chains, photosyntheses, and intermediary metabolism. Bacteria are confronted with a variety of iron limitation conditions. At an anaerobic stage, usually under reducing conditions, enough soluble Fe^{2+} is present to cope with the iron demands. At an aerobic stage, under oxidative conditions, Fe^{3+} is present, which at pH 7 is completely insoluble. In addition, in the human host, most iron exists as free heme, heme bound to hemoglobin, hemopexin, and hemoglobin-haptoglobin. The human host fulfills its own iron requirement by synthesizing the proteins transferrin and lactoferrin, which very tightly bind iron. Free iron in equilibrium with these compounds is orders of magnitude below the concentration (~ 0.1 µM) required to sustain bacterial growth. To cope with iron deprivation, under these conditions, bacteria were highly inventive and developed a number of intricate mechanisms to fulfill their iron requirements. In the following, these mechanisms will be discussed because they form the basis of a huge variety of individual solutions that certain bacteria, under their specific environmental conditions, have developed (Cornelis and Andrews 2010). This becomes especially apparent in the microbial synthesis of a great number of siderophores. A compilation of siderophore structures known since 2009 lists more than 279 siderophores (Hider and Kong 2010) and more have been added since then. Siderophores are synthesized and secreted into the medium where they solubilize the $(Fe^{3+}OOH)_n$ precipitate and transport the Fe^{3+} sidero-phore complex into the cells, where Fe^{3+} is released by reduction and enters the metabolism. In heme transport and transport of iron delivered by transferrin and lactoferrin, siderophores are usually not involved (see Chap. 3).

This chapter only touches iron transport and iron regulatory systems of *Pseudomonas* which are described in depth in a separate chapter in this book.

2.2 Iron Transport into Gram-Negative Bacteria

In Gram-negative bacteria, iron must be transported across the outer membrane and the cytoplasmic membrane. Both transport processes occur independently of each other (Fig. 2.1).

2.2.1 Transport Across the Outer Membrane

Transport across the outer membrane involves transport proteins with 22 anti-parallel β strands that form a β barrel which is tightly closed by a globular plug domain (Noinaj et al. 2010). Regardless of whether iron is delivered by

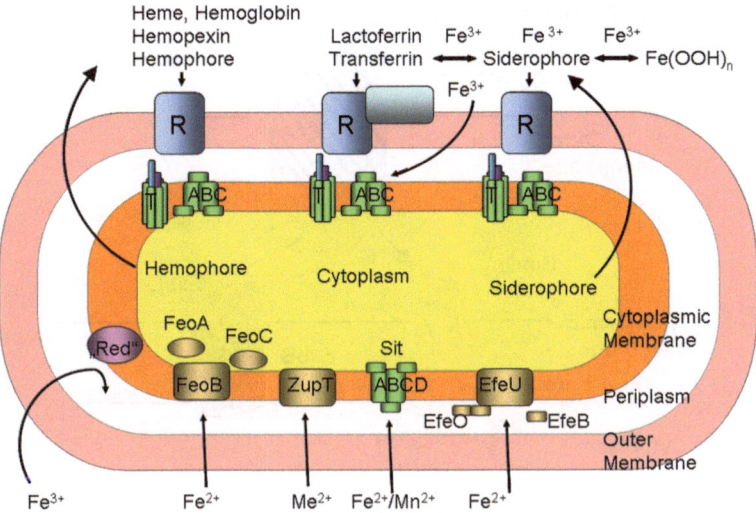

Fig. 2.1 Transport systems of Fe^{2+} and Fe^{3+} into Gram-negative bacteria. R denotes receptor proteins; ABC denotes ATP-binding cassette transporter, T indicates the Ton complex. The heme proteins release the heme at the cell surface which is taken up into the cytoplasm. Transferrin and lactoferrin release iron at the cell surface which is taken up into the cytoplasm. Fe^{3+} siderophores are taken up into the cytoplasm, but some enter only the periplasm. Fe^{2+} diffuses across the outer membrane and is transported by various systems across the cytoplasmic membrane. Red indicates a reduction of Fe^{3+} to Fe^{2+}. Hemophores and siderophores are synthesized in the cytoplasm and released by specific export systems. See text for details

siderophores, heme, hemoglobin, transferrin, or lactoferrin (Fig. 2.1), the iron ligands bind tightly ($K_m \sim 1$ nM) to surface-exposed regions of the transporters. From there, they must be released by changes in the conformation of the binding site and the plug must move so that the Fe^{3+} siderophores, heme, or Fe^{3+} can move into the periplasm. Energy is required to drive these movements which is not generated in the outer membrane but is derived from the proton motive force (pmf) of the cytoplasmic membrane. Coupling of the outer membrane to the cytoplasmic membrane requires a protein complex which consists of the proteins TonB, ExbB, and ExbD. Their location and transmembrane topology are shown in Fig. 2.2. TonB, ExbB, and ExbD are found in cells at a ratio of 1:7:2 (Higgs et al. 2002), but it is unclear whether the proteins are all in the complex or whether some still exist in a soluble form not yet assembled. Detergent-solubilized ExbB forms a hexamer and copurified ExbD forms an $ExbB_6$-$ExbD_1$ complex. TonB has not been copurified in a stoichiometric ratio to ExbB and ExbD. It is assumed that $ExbB_6$ forms the platform on which the final complex is assembled (Pramanik et al. 2010, 2011). Interaction of the three proteins was shown in various ways: ExbB stabilizes TonB and ExbD, and formaldehyde cross-links TonB dimers, ExbD dimers, TonB-ExbB dimers, and ExbB-ExbD dimers (Postle and Larsen 2007). The yield of cross-linked products is too low to unravel larger oligomers. TonB contains in the hydrophobic transmembrane segment a single charged residue, histidine,

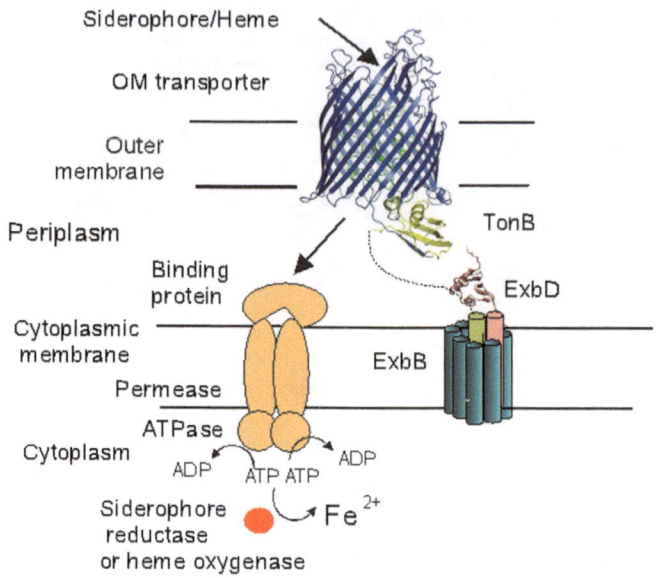

Siderophore/Heme

OM transporter

Outer membrane

Periplasm

TonB

Binding protein

ExbD

Cytoplasmic membrane

ExbB

Permease

ATPase

Cytoplasm

ADP   ATP ATP   ADP

Siderophore reductase or heme oxygenase   Fe^{2+}

Fig. 2.2 Model of siderophore and heme transport across the outer membrane and the cytoplasmic membrane of Gram-negative bacteria. TonB binds to a specific region, TonB box, of the outer membrane transporter which is exposed to the periplasm. The iron compounds are transported across the outer membrane at the expense of energy provided by the proton motive force of the cytoplasmic membrane mediated by the TonB-ExbB-ExbD protein complex. Once in the periplasm, a binding protein binds the iron compounds and delivers them to the ABC transporter in the cytoplasmic membrane. The ABC transporter consists of the transmembrane permease to which an ATPase is attached in the cytoplasm. Fe^{3+} is released from the siderophores by reduction to Fe^{2+} and from heme by heme oxygenase. Heme is also incorporated into heme proteins. See text for details

which was supposed to be the amino acid that responds to the pmf. However, this attractive hypothesis had to be abandoned when it was shown that His20 can be replaced by Asn, while still retaining full TonB activity (Swayne and Postle 2011). Other important amino acids in transmembrane regions are Asp25 in ExbD (Braun et al. 1996) and Glu176 in ExbB (Braun and Herrmann 2004a, b). Asp/Asn and Glu/Ala mutants are completely inactive. These negatively charged amino acids are reasonable candidates for a protonation/deprotonation cycle in response to the pmf across the cytoplasmic membrane, as observed in membrane-bound H^+-ATPases. Protonation/deprotonation may change the conformation of the complex. It is conceivable that TonB in the energy-rich conformation allosterically changes the conformation of the outer membrane transporters. During this process, TonB is "deenergized" and is subsequently reenergized by the pmf to reenter the cycle. TonB physically interacts with outer membrane transporters (Fig. 2.2), as shown by the suppression of point mutants in the N-terminal region (TonB box) of outer membrane transporters by mutations in TonB. The same region was cysteine cross-linked and shown by crystal structures to interact (Pawelek et al. 2006; Shultis et al. 2006). Three early stages in TonB energization were revealed by proteinase K sensitivity and formaldehyde cross-linking assays (Ollis and Postle 2012). To access TonB with proteinase K, spheroplasts were used in which TonB and ExbD were cross-linked depending on the proton motive force, as is the case in cells. TonB was degraded by proteinase K in mutants lacking ExbD, expressing an inactive ExbD(L132Q) or TonB(H20A). No formaldehyde cross-links were formed, which was also the case in an ExbD(D25N) mutant. However, in the latter case, TonB was degraded by proteinase K. In wild-type spheroplasts, TonB was degraded by proteinase K and cross-linked with formaldehyde. In spite of these data, it remains unclear how TonB and the entire complex react to the pmf and how they interact with the outer membrane transporters in a way that changes their conformation, releases the bound substrates, and opens the pore.

A bioinformatic analysis reveals that all Gram-negative bacteria contain one or up to three TonB proteins (Chu et al. 2007). The carboxy-terminal region with which TonB interacts with the outer membrane transporters and receptors assumes a conserved fold. They are either redundant in that they can replace each other or they interact specifically with a transporter. In Vibrio strains, TonB1 is associated with ExbB1 and ExbD1, TonB2 is associated with ExbB2 and ExbD2, and TonB3 is associated with ExbB3 and ExbD3. In addition, vibrios contain TtpC proteins associated with the second and third TonB system. TtpC is a 49 kDa protein that is predicted to span the cytoplasmic membrane three times with its C-terminus in the cytoplasm and the majority of the protein in the periplasm. TtpC is essential for TonB-mediated iron transport systems; however, its mode of action has not been resolved. TonB2 and TonB3 are shorter than TonB1, which does not need a TtpC. It has been hypothesized that TtpC is required for the shorter TonBs to contact the outer membrane transporters (Kustusch et al. 2011).

2.2.2 FhuA Transports Ferrichrome

FhuA is a multifunctional transporter that not only transports ferrichrome across the outer membrane, but also the structurally similar antibiotic albomycin, the structurally diverse antibiotic CGP4832, a synthetic rifamycin derivative, the peptide antibiotic microcin J25, and colicin M. It also serves as a receptor of the phages T1, T5, and Φ80. For all of these functions, FhuA is essential. FhuA of Salmonella strains also transports ferrichrome and albomycin, but the phage receptor specificity differs (Killmann et al. 1998). FhuA was the first outer membrane transport protein whose crystal structure was determined (Ferguson et al. 1998; Locher et al. 1998). Its basic design is exemplary for all of the other outer membrane transporters whose crystal structures were solved (Table 2.1).

Table 2.1 Crystal structures of TonB-dependent outer membrane transporters

Protein	Ligand[a]	PDB ID
FhuA, *E. coli*	–	1BY5
	Ferrichrome	1FCP
	Ferrichrome/LPS	2FCP
	CGP4832/LPS	1F11
	Albomycin	1QKC
	TonB/ferricrocin	2GRX
FecA, *E. coli*	–	1KMO
	–	1PNZ
	Diferric dicitrate	1KMP
	Diferric dicitrate	1PO3
	Dicitrate	1POo
FepA, *E. coli*	–	1FEP
Cir, *E. coli*		2HDF
BtuB, *E. coli*		1NQE
	TonB, vitamin B_{12}	2GSK
	Ca^{2+}	1NQG
	Ca^{2+} vitamin B_{12}	1NQH
FpvA, *P aeruginosa*	–	2O5P
	pyoverdine	2W16
	Ferric pyoverdine	2IAH
	Ferric pyoverdine	1XKH
FptA, *P. aeruginosa*	Pyochelin	1XKW
FauA, *B. pertussis*	–	3EFM
HasR, *S. marcescens*	HasA/heme	3CSL
	HasA	3CSN
ShuA, *S.dysenteriae*		3FHH
TbpA, *N. meningitidis*	Human transferrin	3V8X
FyuA, *Y. pestis*	Yersiniabactin	–
YiuR, *Y. pestis*		–

[a] Not all structures and ligands are listed. The *Yersinia pestis* FyuA protein was recently published (Lukacik et al. 2012), YiuR was not published (Noinaj et al. 2010)

2.2.2.1 Reconstitution of FhuA

FhuA was in vivo reconstituted from the β barrel domain and the plug domain, both equipped with a signal sequence for secretion across the cytoplasmic membrane (Braun et al. 2003a, b). The yield of active protein was rather high (45 %). This suggests that the β barrel folds independent of the plug and that the plug is inserted after the β barrel has formed. It is not clear whether folding is completed in the periplasm or whether final stages occur during insertion into the outer membrane. To yield an active FhuA, no precise plug structure (residues 1–160) is required as a much larger fragment (residues 1–357) also forms an active FhuA.

FhuA has also been reconstituted in a functional state in planar lipid bilayer membranes (Udho et al. 2009). FhuA does not conduct ions across planar lipid bilayer membranes (Braun et al. 2002a, b), since the plug tightly closes the pore in the β barrel. However, the addition of 4 M urea to the *cis* compartment to which FhuA was added results in a continuous rise of conductance for tens of minutes with occasional single channel events of 0.1–1 nS. Upon removal of urea, conductance through FhuA fell 60–90 % and could be repeated several times by the addition and removal of urea. The addition of 4 M urea or 3 M glycerol to the *trans* solution abolished conductance which was established upon removal of urea or glycerol from the trans solution. Urea exerts two functions: it unfolds the plug and it establishes an osmotic pressure gradient across the membrane. If phage T5 was added to either the *cis* or the *trans* solution, no channels were formed. However, when 3 M glycerol was added to the *cis* solution and subsequently perfused out, T5 added to the *cis* or the *trans* solution caused the appearance of numerous 0.1–1nS channels. The glycerol gradient oriented FhuA in the membrane such that phage binding sites faced the *cis* as well as the *trans* compartment. Ferrichrome added to the *trans* side stopped and partially reversed the 4 M urea-induced conductance of FhuA, showing that the ferrichrome binding site was intact in the reconstituted FhuA. To some extent, urea mimics the TonB-dependent reorientation of the plug within the barrel or movement of the plug out of the barrel.

2.2.3 FepA Transports Fe^{3+} Enterobactin

FepA transports cyclic Fe^{3+} enterobactin and its linear forms. It also serves as a receptor for colicins B and D. Fe^{3+} enterobactin transport by FepA has been studied in most detail and unraveled the essential parameters of all these types of transporters (Newton et al. 2010). Although the number of FepA molecules in fully induced cells (35,000 per cell) far exceeds the number of TonB molecules per cell (1000), all FepA molecules are engaged in transport with a maximum turnover number of approximately 5/min and an activation energy of 33-35 kcal/mol. Accumulation of Fe^{3+} enterobactin in the periplasm requires the FepB periplasmic binding protein. In the absence of FepB Fe^{3+}, enterobactin escapes through TolC into the medium. FepB does not interact with FepA. FepA was saturated with Fe^{3+} enterobactin with a K_D of 0.2 nM. Simultaneous transport of Fe^{3+} enterobactin via FepA and ferrichrome via FhuA does not change the K_m and V_{max} of Fe^{3+} enterobactin transport, but reduces the V_{max} of ferrichrome transport by ~ 50 %. TonB does not seem to be a decisive limiting factor in a simultaneous cotransport of various Fe^{3+} siderophores. The rather high activation energy presumably results from movement of the plug to get Fe^{3+} enterobactin through the FepA pore. The intracellular iron concentration has been determined to be in the range of 0.1 to 0.25 mg/g dry weight, which translates into 0.66–1.7 mM or 0.6–1.6 $\times 10^6$ ions per cell. If this quantum is transported per cell cycle, it would cost 2.4–6 $\times 10^6$ ATP equivalents.

E. coli expresses additional outer membrane transporters, Cir, Fiu, and IroN, which recognize with different affinities the iron complexes of cyclic enterobactin, its linear forms (2,3 dihydroxybenzoylserine)$_{1-3}$, and salmochelin and its linear forms. They also bind microcins with catecholate moieties (Müller et al. 2009).

2.2.4 FecA Transports Ferric Citrate and Elicits a Transcription Regulatory Signal

A TonB-dependent iron transport system exists in *E. coli* that is mediated by citrate. Determination of the crystal structure revealed diferric dicitrate bound in a cavity at the cell surface (Ferguson et al. 2002; Yue et al. 2003). Comparison with the unloaded FecA protein reveals a strong movement of two surface loops so that they occlude the entry of the cavity. If ferric citrate is released from the FecA binding site by interaction of FecA with energized TonB, it cannot escape into the medium but diffuses vectorially across the opened pore in FecA into the periplasm. Since citrate is not cotransported with iron into the cytoplasm, it is likely that iron is transported across the cytoplasmic membrane by the FecBCDE ABC transporter (see also Sect. 6.1). FecA was the first case where it was shown that it not only transports ferric citrate, but initiates the transcription of the *fecABCDE* transport genes. Binding of ferric citrate to FecA elicits a signal that is conferred across the outer membrane to the FecR regulatory protein in the cytoplasm which transfers the signal across the cytoplasmic membrane to the FecI sigma factor. In contrast to outer membrane transporters which do not regulate transcription, FecA and all signaling transporters contain a N-terminal extension that extends into the periplasm and serves to interact with FecR (summarized in Braun 2010; Braun et al. 2003a, b, 2006; Braun and Mahren 2005, 2007; Ferguson et al. 2007; Noinaj et al. 2010). A genome survey identified the presence of signal receptors/transducers of the FecA type in a variety of Gram-negative bacteria (Koebnik 2005).

2.2.5 TbpA/TbpB of Neisseria meningitidis Transport Transferrin Iron

Neisseria do not synthesize siderophores but they can use siderophores of other organisms, called xenosiderophores, such as aerobactin, enterobactin, salmochelin, a diglycosylated derivative of enterobactin (Bister et al. 2004), and dimers and trimers of dihydroxybenzoylserine which are derived from enterobactin. Their major iron source is human transferrin that binds to the receptor TbpA and its coreceptor TbpB (Fig. 2.1). TbpA is a rather large outer membrane protein of 100 kDa that belongs to the TonB-dependent transporters. The structure of TbpB differs from the structure of TbpA (Moraes et al. 2009), it is smaller (80 kDa), and it is attached by a lipid anchor to the outer membrane. TbpA binds iron-loaded and unloaded transferrin with similar affinities, whereas TbpB binds only iron-loaded transferrin. TbpA is sufficient to transport iron but iron uptake is more efficient in the presence of TbpB. To unravel how tightly bound iron ($K_a = 10^{23}$ M^{-1}) is transferred from transferrin to TbpA, the crystal structure of TbpA with unloaded transferrin was determined (Noinaj et al. 2012). A special feature of the structure is a long plug loop that protrudes ~ 25 Å above the cell surface, whereas in the other transporters the plug is buried in the β barrel. With this plug loop, TbpA interacts with the C1 subdomain of transferrin. This interaction induces a partial opening of the cleft in the transferrin C-lobe that destabilizes the iron coordination site, and thereby facilitates the release of iron and its transfer to TbpA. In addition, the α-helix of the extracellular loop 3 of TbpA is inserted into the cleft between the C1 and C3 subdomains of the transferrin C-lobe. The latter interface contains residues found only in human transferrin which may explain the specificity of Neisseria TbpA for human transferrin. X-ray and SAXS analysis of TbpB with iron-loaded transferrin revealed binding of TbpB to the C-lobe of transferrin, but at sites that differ from the sites TbpA binds to the C-lobe. Iron bound to the N lobe of transferrin does not affect the binding of transferrin to TbpA and TbpB. All three proteins form an enclosed chamber with a volume of ~ 1000 Å^3 which is located directly above the plug domain of TbpA. The released iron is guided to the pore of TbpA and loss by diffusion into the medium is prevented. Steered molecular dynamics were applied to obtain an idea of how iron may diffuse from the cell surface through TbpA into the periplasm. In the ground state, a large, highly negatively charged transmembrane cavity is located between the barrel and the plug whose access is restricted by a loop from the extracellular side and by a helical gate of the plug from the periplasmic side. When force is applied to the plug domain, that might mimic the action of TonB (Gumbart et al. 2007), first the helical gate and then the restriction loop are removed, producing an unobstructed pathway from the extracellular space to the periplasm.

2.2.6 Ferredoxins Enhance Growth of Pectobacterium spp Under Iron-Limiting Conditions

P. carotovorum and *P. atrosepticum* form colicin-type bacteriocins, which consist of two domains, a N-terminal [2Fe-2S] ferredoxin with 60 % identity to spinach ferredoxin and a C-terminal toxin with 46 % identity to the activity domain of colicin M (Grinter et al. 2012). The two toxins, known as pectocin M1 and pectocin M2, display an increased activity against certain *P. carotovorum* and certain *P. atrosepticum* strains when these are grown under iron-limiting conditions, suggesting the synthesis of an outer membrane receptor for the ferredoxin moiety that serves as an iron source. Spinach ferredoxin prevents the cytotoxicity of M1, suggesting competition with binding to the transporter. Growth enhancement and sequence homology is specific for ferredoxins of plants on which the pectobacteria thrive. Until now, the receptor protein has not been identified and no iron transport measurements were performed.

Replacement of Asp222 by Ala, which is equivalent to the Asp225 essential for the activity of colicin M (Helbig and Braun 2011) inactivates M1. Another *P. carotovorum* strain produces a third bacteriocin, named pectocin P that consists of an N-terminal ferredoxin and a C-terminal domain that is 41 % identical to the activity domain of pesticin of *Yersinia pestis* (Grinter et al. 2012). Pesticin—like all colicin-type bacteriocins—consists of three domains of which the N-terminal region was experimentally fused to lysozyme (gene e product) of phage T4, resulting in a hybrid protein that killed cells (Patzer et al. 2012). M1, M2, and pectocin P are further examples of the evolution of bacteriocins by the assembly of DNA fragments that encode receptor binding, translocation, and activity domains (Braun et al. 2002a, b).

2.2.7 Heme Uptake by Bacteria

2.2.7.1 Heme Uptake Via Hemophores: The HasR/HasA System of *Serratia marcescens*

The particular feature of heme uptake by *S. marcescens* is the participation of a secreted protein, named HasA hemophore (Fig. 2.1), which was discovered in studies of type I protein secretion of *S. marcescens* (Létoffé et al. 1994). HasA tightly binds heme ($K_d = 18$ pM) or heme from hemoglobin, leghemoglobin, hemopexin, and myoglobin without forming stable complexes with these proteins (Wandersman 2010; Wandersman and Stojiljkovic 2000) (Fig. 2.1). HasA binds to the TonB-dependent outer membrane transporter HasR in the heme loaded and the heme unloaded form with a K_d of 5 nM. Two large HasR extracellular loops contact HasA. Transfer of heme from its high-affinity site in HasA to the low-affinity site in HasR against a 10^5 affinity gradient is not energy-driven and occurs in vitro. The crystal structure of HasA bound to HasR reveals a possible mechanism of heme transfer from HasA to HasR (Krieg et al. 2009). Ile 671 in an extracellular loop of HasR contacts heme in HasA resulting in the disruption of the HasA-Tyr75 heme coordination. Replacement of Ile 671 by Gly abolishes heme transfer from HasA to HasR. TonB and pmf are required to release heme-free HasA from HasR. A specific TonB variant, HasB, binds to HasR with high affinity (~ 10 nM), which suggests that the two proteins do not dissociate during the energization/denergization cycle of heme transport across HasR and release of HasA from HasR. Hemophore-mediated heme transport has also been found in *Yersinia pestis, Yersinia enterocolitica, Erwinia carotovora, Pseudomonas aeruginosa,* and *Pseudomonas fluorescens.*

2.2.7.2 HxuC/HxuA of *Haemophilus influenzae* Transport Hemopexin Heme

H. influenzae is a heme auxotroph that takes up heme from various sources of its human host such as hemoglobin, hemoglobin/haptoglobin, hemopexin, and serum albumin via several TonB-dependent outer membrane transporters. HuxC/HuxA is required for using heme from hemopexin, which binds heme with extreme affinity (K_d below picomolar). HxuC is a TonB-dependent outer membrane transport protein, HxuA is secreted by HxuB and a part of it, depending on the *H. influenzae* strain, is released into the medium (hemophore). In *E. coli*, 95 % of HuxA are found to be cell-associated. The *hxuABC* genes were cloned in a heme auxotrophic strain of *E. coli* and the transport system reconstituted (Fournier et al. 2011). In addition, the *H. influenzae* TonB-ExbB-ExbD complex is required for efficient hemopexin heme uptake by *E. coli*. Heme acquisition from hemopexin requires the HxuC transporter and HxuA, which could also be added from outside of the cells. The replacement of two cysteine residues by serine increased the amount of HxuA in the supernatant. It was with this derivative that in vitro studies were performed. HxuA formed a 1:1 complex with heme-free and heme-loaded hemopexin; however, the ∆H values differed by -71 kJ mol^{-1}. The binding of HxuA to hemopexin changes the heme environment; no binding of heme to HxuA could be found. The binding of HxuA to hemopexin also releases heme, which can even be transferred to HasR of *S. marcescens*. The occurrence of the Hxu system in most clinical isolates of *H. influenzae* supports the importance of this heme uptake system for heme acquisition.

2.3 TonB-Independent Iron Uptake Across the Outer Membrane

An ABC transporter encoded by the *sfuABC* genes of *Serratia marcescens* transports iron into a TonB, ExbB, and ExbD mutant (Angerer et al. 1990, 1992; Zimmermann et al. 1989) (Fig. 2.1). It was the first of this category that was characterized. No siderophores and no outer membrane proteins could be related to this iron transport system. FeCl$_3$ and Fe^{3+} citrate supported growth of an *E. coli* K-12 mutant devoid of enterobactin synthesis that had been transformed with the *sfuABC* genes. In strong iron deficiency conditions, growth was stimulated by citrate, which was dependent on active transport by the FecA outer membrane receptor and TonB. Under these conditions, the diffusion of iron across the outer membrane was no longer sufficiently fast to support growth. However, SfuABC transported iron delivered by citrate across the cytoplasmic membrane in a *fecCDE* mutant required for citrate-mediated iron transport in *E. coli*. Apparently, iron was dissociated from citrate and was taken over by SfuABC. Although the experiments were performed with ferric iron under aerobic conditions, the possibility that ferric iron was reduced to ferrous iron prior to uptake across the outer membrane and the cytoplasmic membrane cannot be excluded. A very similar system was described for all *Yersinia* species pathogenic for humans, called YfuABC (Saken et al. 2000). *Haemophilus influenzae* contains an ABC transporter for iron which does not seem to be coupled to a TonB-dependent outer membrane transporter (Adhikari et al. 1995).

In the *Neisseria gonorrhoeae* strain FA19, the FbpABC proteins are required for ferric iron transport removed from transferrin and lactoferrin through the periplasm and across the cytoplasmic membrane. This system also transports iron by enterobactin, its linear monomeric product dihydroxybenzoylserine and the linear form of salmochelin designated S2, each provided in 10 μM concentrations (Strange et al. 2011). *N. gonorrhoeae* does not produce siderophores, but can use xenosiderophores.

In *Vibrio cholerae,* the VctPDGC ABC cytoplasmic membrane transport system promotes iron uptake by vibriobactin, a catecholate siderophore, but also transports iron independent of a siderophore and TonB (Wyckoff and Payne 2011).

In *Haemophilus influenzae*, iron is taken up by the HitABC transporter (Adhikari et al. 1995). No TonB-dependent outer membrane transporter was related to this system.

Legionella pneumophila forms the siderophore legiobactin, which is taken up through the outer membrane protein LbtU. *In silico* analysis of LbtU reveals a typical outer membrane protein with long surface loops and short periplasmic turns. However, in contrast to the 22-stranded ß barrels of TonB-dependent transporters, LbtU consists of only 16 ß strands. *L. pneumophila* does not contain a TonB protein. Therefore, the iron complex of legiobactin is taken up across the outer membrane via LbtU in a TonB-independent manner (Chatfield et al. 2011).

2.4 Hemophore-Type Heme Uptake in Gram-Positive Bacteria

The former assumption that heme diffuses across the cell wall of Gram-positive bacteria until it reaches the cytoplasmic membrane, where the heme transport systems reside, suffers from the fact that free heme is rather insoluble and usually very tightly bound to proteins such as hemoglobin, hemoglobin/haptoglobin, and hemopexin. To cope with this situation, Gram-positive bacteria express intricate hemophore-type heme acquisition systems (Fig. 2.3).

2.4.1 Heme Uptake into Staphylococcus aureus

S. aureus, in which most of the original investigations have been performed (Mazmanian et al. 2003; Hammer and Skaar 2011), transports heme vectorially

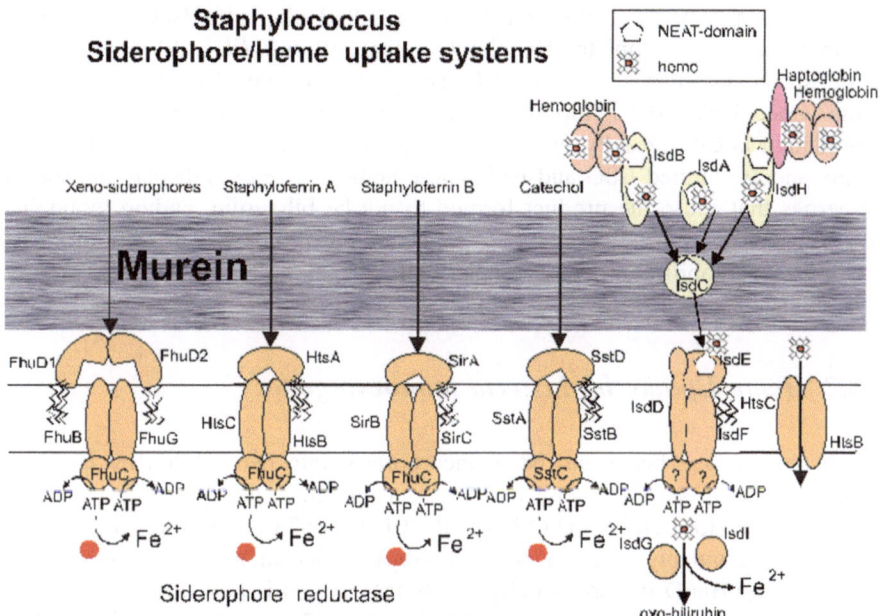

Fig. 2.3 Fe^{3+} siderophore and heme transport into *S. aureus* as an example for Gram-positive bacteria. Fe^{3+} siderophores and Fe^{3+} xenosiderophores are supposed to diffuse unspecifically across the cell wall composed of the murein, wall teichoic acids, teichuronic acids, and proteins to the cytoplasmic membrane, through which they are transported by specific ABC transport systems. In contrast, for heme, a specific diffusion cascade exists through the cell wall along the indicated proteins. In the cytoplasm, Fe^{3+} is mobilized from the siderophores by reduction to Fe^{2+} and from heme by conversion to oxobilirubin. See text for details

across the thick cell wall by a cascade of proteins which are thought to be positioned such that heme moves within single proteins and from one protein to the next protein until it reaches the cytoplasmic membrane. A portion of these proteins is released into the medium like the hemophores of Gram-negative bacteria. The proteins are designated Isd from *i*ron-regulated *s*urface *d*eterminant. They are covalently anchored to the peptidoglycan by sortase A, except IsdC, which is anchored by sortase B. They contain up to three structurally conserved NEAT domains of 125 amino acids that bind heme. Transcription of the *isd* genes is controlled by iron through a Fur repressor. From a number of in vivo and in vitro studies, the following model of heme uptake has been derived. Heme from hemoglobin-haptoglobin is mobilized by NEAT 1 of IsdH and then transferred to NEAT 2 and NEAT 3 of IsdH (Fig. 2.3). Heme from hemoglobin is mobilized by NEAT1 of IsdB and transferred to NEAT 2 of IsdB. Heme is then transferred from IsdH and IsdB either directly to the IsdE binding protein of the transport system in the cytoplasmic membrane, or through IsdA and IsdC. IsdE then transfers heme to the IsdF permease, which—together with IsdD—transports heme across the cytoplasmic membrane. Heme transfer from IsdH/IsdB, IsdA, IsdC, and IsdE is unidirectional and complete. Knockout mutants of *isdDEF* still show reduced heme uptake because a second ABC transporter encoded by *htsBC* transports heme. Once inside the cytoplasm, iron must be mobilized from heme. Heme degradation is catalyzed by two heme oxygenases which are very similar structurally and are termed IsdG and IsdI. These heme oxygenases display the special properties that cause the product formed to not be biliverdin, carbon monoxide, and iron (which are usually created by heme oxygenases), but a mixture of β- and σ-isomers of oxobilirubin, called staphylobilin (Hammer and Skaar 2011).

2.4.2 Heme Uptake by Listeria monocytogenes

Acquisition of free heme and heme incorporated into hemoglobin by *L. monocytogenes* also involves proteins covalently anchored to the peptidoglycan. They are designated Hbp1 and Hbp2 (Xiao et al. 2011). Hbp2 is the primary acceptor for heme and presumably transfers heme to Hbp1, which then delivers heme to the heme-binding HupD lipoprotein ($K_D = 26$ nM), which is part of the ABC transporter in the cytoplasmic membrane. Hbp1 and Hbp2 are only required when the heme concentration is lower than 50 nM. At higher heme concentrations, enough heme diffuses across the cell wall to be directly transported by the HupDGC transporter into the cytoplasm. The overall K_M for heme uptake is 1 nM with a $V_{max} = 23$ pmol 10^9 cells^{-1} min^{-1}. A second heme uptake system accounts for the residual transport of heme in *hupDGC* mutants.

2.5 Mobilization of Transferrin Iron by Staphyloferrins and Catecholamines

S. aureus grows on transferrin as a sole iron source. No protein cascade as shown for heme uptake and no receptor/transporter as present in *N. meningitidis* are involved in transferrin iron uptake by *S. aureus*. Rather, siderophores mobilize transferrin iron and donate it to siderophore-specific ABC transporters in the cytoplasmic membrane. Staphyloferrin A and B and also host catecholamine hormones support growth of *S. aureus* on iron-loaded transferrin. The latter reduce transferrin Fe^{3+} to Fe^{2+} for which transferrin has a low affinity. Staphyloferrin iron is taken up via the HtsABC and SirABC systems, whereas catecholate iron is taken up by the newly identified SstABCD transport system (Beasley et al. 2011).

2.6 Iron Transport Across the Cytoplasmic Membrane

2.6.1 ABC Transporters

ABC transporters are the prevailing systems for Fe^{3+} siderophores, heme, and iron transport across the cytoplasmic membrane. The name ABC is derived from an ATP-binding cassette, which indicates that an ATPase associated from the cytoplasmic side to the transmembrane transport protein, is an essential component of the transport system. The transport proteins consist of one or two different subunits to which two ATPases are usually bound. The substrates are delivered by binding proteins which are either in Gram-negative bacteria free in the periplasm or in Gram-positive bacteria anchored by a lipid moiety of the murein lipoprotein type to the cytoplasmic membrane. The binding proteins determine the substrate specificity of the transport systems. The iron transport proteins usually consist of 20 transmembrane segments, as has first been determined for FhuB of the *E. coli* ferrichrome transport system (Groeger and Köster 1998). In this chapter, 42 known transport proteins of Fe^{3+} siderophore transport systems were in silico analyzed and shown to be arranged in the cytoplasmic membrane like FhuB. Early mutant analyses suggesting active sites in FhuB, the FhuD periplasmic binding protein, the FhuC ATPase, and the interaction sites between these proteins were summarized by Braun et al. (1998) and Köster (2005). After the transfer of iron, Fe^{3+} siderophores, or heme across the outer membrane by the TonB-dependent transporters, the iron compounds bind to periplasmic binding proteins which deliver the iron compounds to the permease in the cytoplasmic membrane. Based on a considerable number of crystal structures and amino acid sequences, the bacterial binding proteins were ordered in a number of classes (Chu and Vogel 2011). The periplasmic binding proteins for Fe^{3+} siderophores belong to the type III structures, as was first determined with FhuD (Clarke et al. 2000). They adopt two independently folded amino- and carboxy-terminal domains with the particular feature of a single long α-helix of approximately 20 amino acids that connects the two domains. The α-helix restricts movement of the two domains relative to each other as is observed in other binding proteins which undergo a "Venus fly trap" movement when they bind their substrates in the cleft between the two domains. Consequently, the substrate of FhuD remains exposed to the surface; for example, the antibiotic part of albomycin remains flexible and is not seen when it is bound via its Fe^{3+} hydroxamate moiety to FhuD (Clarke et al. 2002). The small structural change observed upon substrate binding raises the question of how substrate-loaded FhuD is recognized by FhuB. Molecular dynamics simulations indicate that the C-terminal domain closes 6° upon substrate release, which is considered to be a sufficiently large change to be seen by FhuB (Krewulak et al. 2005).

The crystal structure of a cytoplasmic membrane component of an ABC transporter related to iron transport has not been determined. The closest relative of iron transporters is the vitamin B_{12} transporter, which in fact was the first ABC transporter that was crystallized and the structure determined (Locher et al. 2002).

Vitamin B_{12} is transported across the outer membrane by the high-affinity TonB-dependent BtuB transporter. In the periplasm, vitamin B_{12} is bound by the BtuF binding protein. The ABC transporter consists of two transmembrane spanning subunits (BtuC) which provide 20 transmembrane helices that form a translocation pathway. Two identical ATPases (BtuD) are bound to BtuC at the inner side of the cytoplasmic membrane. The three proteins form a stable, high-affinity complex ($K_d \sim 10^{-13}$). Vitamin B_{12} accelerates the BtuCD-BtuB complex dissociation rate 10^7-fold, which is further destabilized by ATP (Lewinson et al. 2010). Exposed to the cytoplasm is a large water-filled channel. Based on the crystal structures, quantitative binding determinations and electron spin resonance spectroscopy, the following transport reaction cycle is proposed (Joseph et al. 2011). BtuF loaded with vitamin B_{12} binds to the apo or ADP state of BtuCD. Vitamin B_{12} is released to the translocation channel. ATP binding to the ATPases inwardly opens the translocation channel and vitamin B_{12} diffuses into the cytoplasm. Concomitantly, the periplasmic BtuC gate is closed. Excess vitamin B_{12} in the periplasm and ATP bound to BtuD promotes dissociation of BtuF from BtuCD and ATP is hydrolyzed. BtuCD is restored to an outward-facing conformation, ready to interact with another vitamin B_{12}-loaded BtuF. The mechanism proposed for vitamin B_{12} transport across the cytoplasmic membrane may apply to most iron transport systems.

The mechanism of vitamin B_{12} transport differs from the mechanism of maltose transport, which is the model system for bacterial type I ABC transporters. These importers are composed of 10–14 helices and alternate from an ATP-bound outward-facing conformation where the periplasmic binding protein releases its substrate to the low-affinity binding site of the transmembrane protein, to an ADP-bound inward-facing conformation where the substrate is released into the cytoplasm. In the crystal structure, in the absence of maltose binding protein, MalFGK$_2$ (MalFG transmembrane proteins, MalK ATPase) forms an inward-facing conformation with the transmembrane maltose binding site exposed to the cytoplasm. The crystal structure MalFGK$_2$ with unloaded maltose binding protein and ATP shows that closure of the nucleotide binding domains of MalK is concomitant with the transfer of maltose from the binding protein to the MalFG transmembrane domains. Interactions of maltose-loaded binding protein induce partial closure of the MalK dimer in the cytoplasm. ATP binding to this conformation promotes progression to the outward-facing state (Oldham and Chen 2011).

2.7 Transport of Fe^{2+}

2.7.1 The Feo System

Under anaerobic/microaerophilic conditions and/or low pH, iron may occur in the Fe^{2+} form, which has a much higher solubility at pH 7 (0.1 M) than Fe^{3+} (10^{-18} M). In 1987, an Fe^{2+} transport system was discovered in *E. coli* (Hantke 1987) and in the meantime found widely distributed in bacterial genomes (Cartron et al. 2006). The *feo* operon consists of three genes—*feoA*, *feoB*, *feoC*—arranged in this order and transcribed from a Fe^{2+}-Fur-regulated promoter upstream of *feoA*. Only FeoB is essential for Fe^{2+} transport. FeoB is an unusual bacterial transport protein that consists of three domains, an N-terminal GTPase (Marlovits et al. 2003) exposed to the cytoplasm which is linked by a spacer to the C-proximal transmembrane domain. The spacer may function as a GDP dissociation inhibitor (Eng et al. 2008). This kind of structure resembles eukaryotic regulatory GTPases. The crystal structures of the cytosolic domains of Feo from *E. coli* (Guilfoyle et al. 2009), *Methanococcus jannaschii* (Köster et al. 2009), and *Thermotoga maritima* (Hattori et al. 2009) were determined. They elucidate the GTP binding site. The GDP dissociation inhibitor and the GTPase form a large interface with a hydrophobic and two polar interactions. GDP binding to the GTPase domain is stabilized by the interaction of the GTPase with the GDP dissociation inhibitor. It is speculated that this represents the "off state" of the FeoB transporter; the GTP bound state would be the "on state." Any model must take into account that the GTP hydrolysis rate is much too slow (50 % hydrolysis after 6 h at 37 °C) to serve as an energy source to drive iron import. In contrast, dissociation of GDP is much too fast to regulate the open/closed state. Therefore, additional proteins are most likely involved in FeoB activity regulation, one of which could be FeoA that contains a SH3 domain (Su et al. 2010) which in eukaryotes is involved in protein–protein interactions. It seems that a GTP/GDP cycle regulates the activity of the transporter which resides in the transmembrane portion that makes up most of the protein. No structural or functional studies have been reported on this portion of FeoB. The energy source for Fe^{2+} transport is another open question. In *Helicobacter pylori,* ATP hydrolysis has been claimed to drive Fe^{2+} transport on the experimental basis that FCCP (a protonophore), DCCD (an ATP synthesis inhibitor), and orthovanadate, an ATP hydrolysis inhibitor, abolish high-affinity Fe^{2+} uptake (Ks = 0.54 μM) (Velayudhan et al. 2000). However, the structure of FeoB does not resemble any ATPase, nor does it contain an ATP but rather a GTP binding motif.

Fe^{2+} uptake via FeoB is a major pathway for *H. pylori* Fe acquisition as *feoB* mutants are unable to colonize the gastric mucosa of mice, which is a low pH, low-oxygen environment. The virulence of other bacteria also depends on the Feo iron transport system. For example, *feoB* mutants of *E. coli* show an attenuated ability to colonize the mouse intestine (Stojiljkovic et al. 1993). The LD_{50} of a *feoB* mutant of *Streptococcus suis* was tenfold higher than the LD_{50} of the *feoB* wild-type strain (Aranda et al. 2009). Investigation of the role of iron transport for the virulence of a particular strain is hampered by the formation of redundant iron transport systems, such as, for example, enterobactin, salmochelin, yersiniabactin, aerobactin, and the Feo system in extraintestinal pathogenic *E. coli*. To study the contribution of one system, all of the other systems must be inactivated. In the case of the TonB-dependent Fe^{3+} transport systems deletion of TonB is sufficient to evaluate the contribution of iron to virulence.

2.7.2 The EFE System

Under aerobic conditions, *E. coli* and *Shigella* strains express a Fe^{2+} transport system that is encoded by the *efeUOB* operon (Grosse et al. 2006). This system resembles oxidase-dependent iron transporters in yeast, since EfeU is homologous to Ftr1p of yeast. It resides in the cytoplasmic membrane with seven transmembrane helices and two REXXE motifs that are required for iron transport. The periplasmic EfE belongs to the heme peroxidase family and contains a *b*-type heme.

2.7.3 The Sit System

Besides the common Feo system, all Shigella strains and numerous pathogenic Gram-negative bacteria express a second Fe^{2+} transport system (Fisher et al. 2009) composed of the *sitABCD* genes which form an ABC transporter. SitA is a periplasmic binding protein, and SitB is the ATPase that is attached to the SitCD transmembrane proteins. Sit is not very specific for Fe^{2+} but also transports Mn^{2+}(Kehres et al. 2002). Although Sit transports Fe^{2+}in addition to Mn^{2+}, transcription of the *sit* genes in *S. flexneri* is unexpectedly induced under aerobic conditions and is repressed under anaerobic conditions. Sit is an important virulence factor as it enhances growth of *S flexneri* in cultured epithelial cells and for virulence in a mouse lung model.

2.7.4 VciB of Vibrio cholerae

VciB was discovered as a growth stimulatory factor in low-iron medium of *E. coli* and *S. flexneri* transformed with *vciB* (Mey et al. 2008). VciB enhances iron uptake via the Feo or the Sit Fe^{2+} transport systems; it does not catalyze iron uptake by itself. It resides in the cytoplasmic membrane with a large loop in the periplasm. It is not known how it achieves an increase in Fe^{2+} transport; however, the following possibilities are discussed: 1) VciB acts as a Fe^{3+} reductase that increases the local concentration of Fe^{2+} for transport by the Feo or Sit Fe^{2+} transport systems, 2) VciB binds Fe^{2+} and delivers it to the Feo or Sit transport proteins, and 3) VciB stimulates the activities of the transporters. Since VciB would have to interact with Feo and the Sit transporters, lack of protein specificity argues against a direct interaction with the Feo and Sit proteins. VciB does not stimulate Fe^{3+} transport by Fbp of *V. cholerae* or Hit of *Haemophilus influenzae*.

2.8 Iron Uptake of Marine Bacteria

Fe^{3+} concentrations in oceanic surface waters are as low as 0.02–1 nM, with more than 99 % bound to mostly undefined organic ligands. This extremely low concentration is caused by the virtual insolubility of Fe^{3+} at the slightly basic pH of seawater and the low iron input. Iron is a growth-limiting nutrient in seawater for the phytoplankton. To cope with iron limitation, marine bacteria synthesize siderophores and Fe^{3+} siderophore transport systems. Since cultivation of marine bacteria is usually difficult or impossible, data on iron transport genes and siderophore biosynthesis gene clusters are mostly derived from genomic and metagenomic approaches (Hopkinson and Barbeau 2011) and siderophores may be identified without relating them to a particular strain. TonB-dependent outer membrane transporters (TBTs) are predicted for Fe^{3+} hydroxamates, Fe^{3+} catecholates, and heme. Fe^{3+} hydroxamate and heme transporters are found in almost half of the prokaryotes, and Fe^{3+} catecholate transporters are found in about one-third of the genomes. TBTs are not evenly distributed among marine bacteria. On average, 2.5 TBT genes per genome are encountered but many genomes encode more than 30 TBTs. Some marine bacteria may encode up to 117 TonB-dependent transporters, which certainly do not take up only iron compounds but other metal ions, as has been found for Ni^{2+} (Schauer et al. 2007, 2008), Zn^{2+} (Stork et al. 2010), and in particular for sugars (Blanvillain et al. 2007; Bjursell et al. 2006; Eisenbeis et al. 2008; Lohmiller et al. 2008; Neugebauer et al. 2005). Fe^{3+} ABC transporters are found in nearly all members of the *Cyanobacteria, Alphaproteobacteria*, and *Gammaproteobacteria,* but only in one member of the *Bacteroidetes*. In contrast, examination of the genomes of known abundant, free-living marine picocyanobacteria and *Pelagibacter ubique* revealed only very rarely Fe^{3+} siderophore transport genes; they may take up inorganic iron. There is a negative correlation between the occurrence of Fe^{3+} transport systems and the FeoA/FeoB Fe^{2+} transport system. Marine prokaryotes tend to have only one of these systems, but not both. For example, picocyanobacteria have the Fe^{3+} ABC system but no FeoA/FeoB and the *Bacteriodetes* have FeoA/FeoB but no Fe^{3+} ABC transporter, with one exception: *feoA* is more frequently encountered than *feoB* because *feoA* occurs often in multiple copies per genome. This finding is unexpected as studies with *E. coli* and other bacteria assign the Fe^{2+} transport function to FeoB and FeoA is dispensable.

2.9 Single Protein Transporters Across the Cytoplasmic Membrane

In addition to the ABC transporters—which consist of three to four proteins—there exist iron transporters across the cytoplasmic membrane which are composed of only a single protein. These are less well-studied than the ABC transporters. In

P. aeruginosa, Fe^{3+} pyochelin is taken up across the outer membrane by the TonB-dependent FptA transporter and across the cytoplasmic membrane by the FptX protein (Reimmann 2012). Fe^{3+} pyoverdine is transported by *P. aeruginosa* across the outer membrane by the TonB-dependent FpvA transport proteins. In the periplasm, Fe^{3+} is released from pyoverdine by reduction and pyoverdine is secreted unmodified to the extracellular space by the PvdRT-OmpQ efflux pump, which also secretes newly synthesized pyoverdine, which is, in turn, used for the next round of Fe^{3+} transport. Fe^{3+} pyoverdine is not taken up into the cytoplasm (Greenwald et al. 2007). How iron enters the cytoplasm is not known.

2.10 Iron Release from Siderophores in the Cytoplasm

In cytoplasmic fractions of bacteria, reduction of Fe^{3+} siderophores to Fe^{2+} can usually be encountered. However, few studies specifically related reduction to the use of Fe^{3+} siderophores. Evidence that reductases are specifically involved in iron acquisition comes from genomic localization of putative reductase genes adjacent to Fe^{3+} siderophore transports genes, derepression of reductase gene transcription by iron limitation, and substrate specificity. Examples are FhuF of *E. coli* (Matzanke et al. 2004), YqjH of *E. coli* (Miethke et al. 2011a, b), and FchR of *Bacillus halodurans* (Miethke et al. 2011a, b). FhuF is a 2Fe-2S protein that reduces Fe^{3+} hydroxamates; in particular, ferrioxamine B. *fhuF* mutants show reduced growth on plates with ferrioxamine B as the sole source of iron. Although iron uptake via ferrioxamine B, ferrichrome, and coprogen is unimpaired, iron removal is reduced. In contrast, YqjH catalyzes the NADPH-dependent release of iron from a variety of siderophores, most effectively from Fe^{3+} 2,3 dihydroxybenzoylserine which is formed by intracellular hydrolysis of Fe^{3+} enterobactin. Fe^{3+} dicitrate serves as another substrate for YqjH. *fchR* clusters with a ferric citrate-hydroxamate uptake system. It catalyzes iron removal from Fe^{3+} schizokinen, produced by *B. halodurans*, supplied Fe^{3+} dicitrate, Fe^{3+} aerobactin, and ferrichrome. A *fchR* deletion mutant shows a strongly impaired growth but accumulates iron, indicating that the Fe^{3+} siderophores are not efficiently metabolized. In vivo, FchR is inhibited by redox-inert siderophore mimics, supporting Fe^{3+} siderophore-specific reduction which may be a target for future antibiotics.

The iron-free siderophores are degraded, as has been shown for enterobactin (Greenwood and Luke, 1978), modified like ferrichrome which is acetylated in *E. coli* (Hartmann and Braun, 1980) and *P. aeruginosa,* or secreted as pyoverdine (Hannauer et al. 2010a). Cells must get rid of the siderophores to avoid iron sequestration from cell metabolism.

2.11 Secretion of Siderophores

Siderophores synthesized in the cytoplasm are secreted to the external medium where they scavenge iron and transport it into cells. Only a very few studies are devoted to siderophore export. Enterobactin secretion was the first siderophore secretion system to be characterized. Enterobactin is secreted across the cytoplasmic membrane by the EntS protein (Furrer et al. 2002) and across the outer membrane by the TolC protein (Bleuel et al. 2005). In *P. aeruginosa*, the protein PvdE secretes a precursor of pyoverdine, called ferribactin, across the cytoplasmic membrane. The precursor is then further processed to pyoverdine, which is exported by PvdRT-OpmQ across the outer membrane (Hannauer et al. 2010b). Pyoverdine synthesis is a complex, not entirely resolved process. Precursors contain myristic or myristoleic acid (Hannauer et al. 2012), which may attach the siderophore to the membrane and prevent its diffusion in the cytoplasm and the periplasm where the siderophore may bind iron and withdraw it from synthesis of iron-containing redox enzymes. The PvdRT-OpmQ complex is involved in the secretion of imported pyoverdine after the release of iron in the periplasm. PvdR functions as an adapter between PvdT, a predicted inner membrane protein, and OpmQ. PvdRT-OpmQ is an ABC transporter related to the TolC containing ABC transporters of *E. coli*. In the cyanobacterium *Anabaena* PCC 7120, SchE, a putative cytoplasmic membrane protein, has been implicated in HgdD, a TolC homolog, mediated secretion of schizokinen, a siderophore synthesized by *Anabaena* (Nicolaisen et al. 2010).

2.12 Examples of Newly Discovered Siderophores

To the huge variety of known siderophores (Carvalho et al. 2011; Hider and Kong 2010), each year new siderophores are added. A few examples of these new siderophores will be discussed. Streptomycetes frequently produce desferrioxamines but some synthesize in addition catechol-type siderophores. The first discovered example is griseobactin, of which the genes for biosynthesis, secretion, uptake, and degradation have been identified. Sequencing and mutant analysis defined the biosynthesis pathway. The antibiotic was isolated and its siderophore function characterized. It consists of a cyclic and, to a lesser extent, linear trimeric ester of 2,3-dihydroxybenzoyl-arginyl-threonine (Patzer and Braun, 2010). The versatile capacity of *Streptomycetes* to synthesize a large variety of compounds is supported by the finding that the Gram-positive *Streptomyces scabies* 87–22 synthesizes an active *pyochelin* that has hitherto only been found in Gram-negative *Pseudomonas* strains (Seipke et al. 2011).

A unique mixed-type catecholate-hydroxamate siderophore has been isolated from *Rhodococcus jostii* RHA1. It contains an unusual ester bond between an L-δ-N-formyl-δ-N-hydroxyornithine moiety and the side chain of a threonine residue (Bosello et al. 2011).

Siderophore synthesis genes of the nonribosomal peptide type (NRPS) are found in 27 % of the marine prokaryotic genomes and of the NRPS-independent pathway in 10 % of the genomes (Hopkinson and Barbeau 2011). However, assignment to siderophore biosynthesis is not unequivocal, since other secondary metabolites are synthesized by these pathways.

Marine siderophores have the specific properties of being amphiphilic. For example, the acetyl group at the lysine residue in aerobactin found in many pathogenic Gammaproteobacteria is in a marine *Vibrio* species replaced by fatty acids that range in size from C8 to C12 (Gauglitz et al. 2012). This siderophore belongs to the class of citrate-based amphiphilic siderophores. Marine bacteria form a second class of amphiphilic siderophores which are peptide-based (Vraspir and Butler 2009). The amphiphilic siderophores are associated with the bacterial membranes which may prevent loss of siderophores by diffusion into the ocean.

2.13 Therapeutic Use of Siderophores

Currently, there are three siderophores used for therapeutic purposes (Chu and Vogel 2011). For more than 40 years, desferrioxamine B (DFO, Desferal®) (produced by *Streptomyces pilosus*) has been administered to reduce iron overload in patients who suffer from severe anemia caused by the hereditary disease ß-thalassemia. DFO binds Fe^{3+} tightly in serum and is secreted into bile (~ 70 %) and urine (~ 30 %). However, the preferred treatment using DFO, suffers from poor absorption if taken orally, so it must be injected intravenously or infused subcutaneously for many hours. New chemotherapeutics such as deferiprone (tradename Ferriprox) and deferasirox (trade name Exjade) are active when taken orally. Deferiprone requires three daily doses, whereas deferasirox requires only one. Both compounds still require long-term inspection to monitor side effects.

Another therapeutic approach is the use of active siderophore transport systems to bring antibiotics into cells into which they are poorly taken up (Trojan Horse concept). Most antibiotics enter cells by diffusion through the cell envelope for which their structure may not be well suited. In particular, the outer membrane of Gram-negative bacteria forms a permeability barrier that prevents the generation of a sufficiently high intracellular concentration of antibiotics to kill cells. Attempts to transport antibiotics chemically coupled to siderophores carriers into bacteria were experimentally successful, but did not sufficiently meet the therapeutic requirements. One well-studied example is albomycin, which is composed of a ferric hydroxamate carrier to which the antibiotic thioribosyl pyrimidine is attached. Albomycin is synthesized by the *Streptomyces* species and is taken up by an energy-coupled transport across the outer membrane and the cytoplasmic membrane of bacteria that contain a transport system for ferrichrome. This applies to most Enterobacteriaceae, *Staphylococcus aureus,* and *Streptococcus pneumoniae* (Pramanik and Braun 2006; Pramanik et al.,2007). Inside cells, the antibiotic must be released from the siderophore carrier by peptidases in order to be active (Braun et al. 1983). In a screening for antibiotics that interfere with RNA metabolism, the antibiotic moiety of albomycin was isolated and shown to inhibit in vitro t-RNA synthetases at a minimal inhibitory concentration (MIC_{50}) of 8 ng/ml (Stefanska et al. 2000). This is also the MIC_{50} of albomycin against cells, in contrast to the 256 mg/ml required of the antibiotic uncoupled from the siderophores carrier. In contrast to diffusion of the antibiotic, energy-coupled transport of the siderophores-antibiotic conjugate reduces the MIC_{50} 32,000-fold.

Another antibiotic, CGP 4832, a semisynthetic rifamycin derivative, is >200-fold more active than rifamycin against *E. coli* and *Salmonella,* since it is taken up across the outer membrane by the energy-coupled ferrichrome transport system (Pugsley et al. 1987). Although it differs structurally from ferrichrome and albomycin (Ferguson et al. 2001), it binds at the same site to the outer membrane FhuA transporter (Ferguson et al. 2000). The ferrichrome transport system across the cytoplasmic membrane does not accept CGP 4832. It is sufficient to overcome the permeability barrier of the outer membrane to strongly reduce the MIC_{50}.

Antibiotics with catechol siderophores occur naturally, e.g. certain microcins, and have been chemically synthesized. They exhibit a 100–10,000-fold higher activity than the antibiotics (compiled in Braun 1999; Braun et al. 2009; Ji et al. 2012). To render *E. coli* more susceptible to the aminocoumarin antibiotic clorobiocin, a gyrase inhibitor, a 3,4-dihydroxybenzoyl moiety was added by modifying the biosynthetic gene cluster. The resulting adduct was taken up via catechol siderophore transporters (Alt et al. 2011).

2.14 Host Defense Mechanisms, Nutritional Immunity

Bacterial hosts bind iron to proteins for several reasons. Iron binding proteins serve to establish iron homeostasis, solubilize Fe^{3+}, serve as iron carriers, prevent concentrations of soluble iron that are too high and which would elicit the formation of oxygen radicals that damage DNA, proteins, and membranes, and withdraw iron from bacteria to control their multiplication. The proteins that serve these purposes are transferrin, lactoferrin, ferritin, hemoglobin, haptoglobin, hemopexin, and siderocalin (NGAL lipocalin). Iron withdrawal by these proteins, also called nutritional immunity, is very effective for bacteria that are unable to mobilize the iron sources deposited in these proteins. However, as described in previous sections, commensal and pathogenic bacteria have developed systems with which they can cope with iron limitation. They withdraw the extremely tightly bound iron from transferrin and lactoferrin ($K_D = 10^{-23}$ M^{-1}), and they mobilize heme from hemoglobin, hemopexin, and the hemoglobin-haptoglobin complex. Although bacteria use the protein-bound iron sources, the constant fight for iron limits their proliferation. For example, only 30–40 % of the transferrin molecules are loaded with iron, leaving a substantial capacity to withdraw various forms of iron from bacteria. Lactoferrin is released from neutrophil granules at sites of inflammation, inhibiting the growth of infecting pathogens by directly sequestering iron. It also prevents *P. aeruginosa* from forming biofilms by stimulating surface mobility and renders them more susceptible to endogenous and exogeneous antimicrobials in the planktonic state (Singh et al. 2002).

Another protein that is secreted by host cells at sites of infection is siderocalin, which contains bound enterobactin when it is expressed in *E. coli*. This accidental observation resulted in intense investigation of the occurrence and the binding properties of siderocalins (reviewed by Correnti and Strong, 2012). Siderocalins often bind siderophores with subnanomolar affinities. Scn (also known as NGAL) binds enterobactin-like siderophores and carboxymycobactin. Scn knock-out mice are more susceptible to infections with bacteria that use these siderophores for iron acquisition. Interestingly, diglucosylated enterobactin known as salmochelin, that has been first found in *Salmonella enterica* (Bister et al. 2004) is not bound by siderocalin (Fischbach et al. 2006; Floh et al. 2004; Valdebenito et al. 2007). *S. enterica* and other bacteria avoid growth inhibition by iron deficiency exerted by siderocalin in that they form the glucosylated salmochelin.

The binding of vibriobactin, a siderophore of the catechol type, to the ferric vibriobactin periplasmic binding protein of *Vibrio cholerae* differs from the binding of enterobactin to transport proteins. The three catechol moieties donate five, rather than six, oxygen atoms as iron ligands. The sixth iron ligand is provided by a nitrogen atom from the second oxazoline ring. This particular iron coordination may explain the failure of vibriobactin to bind to siderocalin, thus evading the siderocalin-mediated innate immune response (Li et al. 2012).

Bacterial infections elicit an increased biosynthesis of the hepcidin peptide hormone in the liver that is released into the serum. It inhibits iron uptake by the gut and prevents the release of iron from iron stores in the liver and in macrophages into the circulatory system. The resulting iron deprivation reduces bacterial multiplication to a level that the innate and acquired immune system can eradicate the infection (Ganz 2006).

2.15 Regulation of Bacterial Iron Homeostasis

2.15.1 The Fur Protein Family

The first function observed for the Fur protein in *E. coli* was ferric iron uptake regulation. Fur, loaded with Fe^{2+}, repressed transcription of iron uptake genes. This general regulatory activity of Fur is crucial for many bacteria to achieve iron homeostasis in the cell. Fur regulates a variety of cellular functions which are directly or indirectly related to iron metabolism. For example, in *Anabaena* sp PCC 7120, Fur is essential as no *fur* mutants were isolated. Proteomic analysis revealed Fur targets that belonged to photosynthesis, energy metabolism, redox regulation, oxidative stress, and signal transduction, and thus link these processes to iron metabolism (Gonzales et al. 2011). Proteins of the Fur family are also involved in regulating the uptake of other divalent cations like zinc (Zur) (Patzer and Hantke 1998), manganese (Mur) (Diaz-Mireles et al. 2004; An et al. 2009), and nickel (Nur) (An et al. 2009). PerR of *Bacillus subtilis* is also a member of the Fur family. Its major task is the response to oxidative stress (Faulkner et al. 2012). A special heme-dependent variant of Fur is the Irr protein, which regulates via heme iron homeostasis in *Bradyrhizobium* (Small et al. 2009). The various Fur homologs make it difficult to ascribe a function that predicts Fur proteins in a newly sequenced organism without further physiological studies. Moreover, as observed for many general regulators, Fur proteins also have the opposite functions as they not only repress gene transcription, but directly or indirectly activate genes.

The Fur and Fur-like proteins have an N-terminal winged helix-turn-helix DNA binding domain and a C-terminal domain responsible for dimerization and binding of divalent metal ions. The crystal structure of the Fur protein from Pseudomonas was the first to be published (Pohl et al. 2003). In the meantime, the structures of the *Vibrio cholerae* Fur, *Mycobacterium tuberculosis* Zur, the *Bacillus subtilis* PerR, and the *Streptomyces coelicolor* Nur reveal a similar overall structure of this protein family. The iron binding site has been a matter of debate as discussed in the recent paper on the *Helicobacter pylori* Fur (Dian et al. 2011) and in a recent review on iron-containing transcription factors (Fleischhacker and Kiley 2011). It seems that the metal binding site 2, which connects the C-terminal dimerization domain with the N-terminal DNA binding domain is the iron binding domain (Dian et al. 2011)

2.15.2 General Regulators of Iron Metabolism Other Than Fur

In most bacteria, the Fur protein has been described as a major regulator of iron metabolism. However, in certain bacterial groups, other regulators have taken over this essential function.

In the Rhizobiales and the Rhodobacterales, both belonging to the alphaproteobacteria, the RirA protein operates as the general iron regulator and Fur homologs like Irr, Mur, and Zur have more specialized regulatory functions (Johnston et al. 2007). The RirA protein is a member of the Rrf2 proteins which contain an Fe–S cluster. The Rrf2 family has very divergent functions in regulating the response to NO (e.g. NsrR in *Neisseria*), cysteine metabolism (e.g. CymR in *Bacillus*), and Fe–S cluster biogenesis (e.g. IscR in *E. coli*).

RirA does not dominate iron regulation in all alphaproteobacteria. For example, in *Caulobacter*, another member of the alphaproteobacteria, there is evidence for a Fur protein as general iron regulator (da Silva Neto et al. 2009).

Firmicutes, the low GC Gram-positive bacteria (e.g. *Bacillus* species, *Staphylococcus* species, and others) use Fur as a general iron regulator (Ollinger et al. 2006), whereas in the high GC Gram-positive bacteria, Corynebacteria, Mycobacteria, and Streptomycetes, iron metabolism is regulated by DtxR (diphtheria toxin regulator) (Wennerhold and Bott 2006; Günter-Seeboth and Schupp 1995). DtxR has an overall architecture similar to Fur, although its sequence is not homologous to Fur.

Like Fur, DtxR consists of an N-terminal DNA binding site with a helix-turn-helix motif and a C-terminal dimerization and metal binding domain. In other bacteria, MntR, a DtxR homolog, regulates manganese uptake (Patzer and Hantke 2001).

In addition to DtxR, four Fur homologs are found in *Streptomyces coelicolor*. One of them functions as a Nur nickle regulator (Ahn et al. 2006), and another one as a Zur zinc regulator (Shin et al. 2007).

In the phylum Spirochaetes, some species like *Treponema pallidum* and the *Borrelia* group do not encode Fur homologs. The TroR protein, a member of the DtxR family, is suggested to regulate iron and manganese levels in *T. pallidum* (Brett et al. 2008). However, another member of this phylum, *Leptospira denticola*, has four Fur homologs, but it is not clear if one or some of them are regulating iron transport (Louvel et al. 2006). A more recent publication suggests a role in oxidative stress response for one of these genes (Lo et al. 2010).

A DtxR homolog has also been found in *Deinococcus radiodurans* (Deinococcus thermus group) to inhibit manganese uptake and stimulate iron uptake.

In the cyanobacterium *Anabaena*, FurA regulates iron homeostasis (Gonzalez et al. 2011) and in cyanobacterial genomes, up to four members of the Fur protein family are observed. The toxin microcystin is regulated among others by Fur in the freshwater cyanobacterium *Microcystis aeruginosa*.

A *fur* homolog was seen in the genome of *Thermococcus kodakaraensis*; however, the gene seemed not to be expressed due to a frameshift mutation. A DtxR homolog may regulate in this organism iron supply (Louvel et al. 2009).

Thermotoga maritima and *Chloroflexus aurantiaca* contain Fur homologs; however, nothing is known about their function.

Very little is known about the regulation of iron homeostasis in archaea except *Halobacterium salinarum* in which two DtxR-like proteins Idr1 and Idr2 take care of iron homeostasis. Gene expression profiles and DNA binding of the regulators were analyzed under iron-rich and iron-limiting conditions. These data were used to define the regulons of Idr1 and Idr2. Both regulators bind mainly in the presence of iron to DNA and many of the affected genes are involved in iron metabolism. A certain subset of genes was regulated in concert by Idr1 and Idr2 (Schmid et al. 2011).

2.15.3 Indirect Regulation by Fur

2.15.3.1 Small RNAs

A major principle of indirect regulation by Fur is exerted by small RNAs. In *E. coli*, the small RNA (sRNA) RyhB is 90 nucleotides long and its synthesis is repressed by Fe^{2+}Fur. In iron deficiency, the production of this RNA is derepressed and about 20 genes are downregulated. With Hfq as a cofactor for the pairing of sRNA and mRNA, the degradation of RhyB and the bound mRNA is triggered. The products of the regulated genes often need high amounts of iron for maturation as e.g. succinate dehydrogenase. Their downregulation helps the cell to spare iron. However, it has also been observed that a few genes are upregulated (Prevost et al. 2007). For example, *shiA* encoding a shikimate transporter is upregulated by RhyB. The induction of *shiA* may help the cells to obtain more of the precursor for the synthesis of enterobactin, the siderophore which allows a better iron supply. In addition, RyhB is necessary for good expression of the enterobactin biosynthesis gene cluster *entCEBAH* (Salvail et al. 2010). The amount of active Hfq oligomers is regulated by the stringent response regulator RelA. The production of ppGpp by RelA seems not to be involved in the regulation of Hfq activity (Argaman et al. 2012). However, ppGpp production by SpoT is increased in iron deficiency and this high ppGpp level stimulates the expression of iron uptake genes in an unknown way (Vinella et al. 2005). RhyB is also involved in the regulation of the virulence factor IscP in *Shigella* (Africa et al. 2011). IscP is an outer membrane protease which degrades IscA, the outer membrane protein responsible for actin polymerization on the surface of the bacterium. This polymerization results in an actin tail which moves the bacterium within cells and into neighboring cells. This regulation allows the bacterium to lower the synthesis of IscP inside colonic cells where the available iron is low and actin-based movement aids infection.

2.15.3.2 ECF Sigma Factors

Bacteria are able to respond specifically to the presence of certain siderophores in their environment by the induction of specific transport systems. Recognition of these siderophores is often dependent on an extracytoplasmic function (ECF) sigma factor. In the Fec system of *E. coli*, the sigma factor is called FecI and the cognate signaling protein FecR (Braun and Mahren 2005). The expression of FecI and FecR is regulated by ferric citrate iron and Fur. The outer membrane transporter FecA has to bind the substrate Fe^{3+}-dicitrate and in a TonB-dependent manner, this binding is recognized by a structural change of FecR, which interacts with an N-terminal domain of the transporter. The structural change of FecR, which is anchored by one transmembrane helix in the cytoplasmic membrane, leads to proteolytic degradation in the cytoplasmic membrane by the site-2 protease RseP. The resulting cytoplasmic FecR fragment binds to FecI which is activated. Active FecI recruits the RNA-polymerase to the promoter of the *fec* operon. In *Pseudomonas*, several similar systems are found which have been analyzed carefully (Draper et al. 2011). The activation of the ECF sigma factor HurI in *Bordetella bronchiseptica* is also dependent on the predicted site-2 protease HurP—this system regulates the uptake of heme. An N-terminal elongation of the receptor protein by about 80 residues and the presence of a FecIR type pair is a good indication of ECF sigma factor regulation. This type of regulation is found in many Gram-negative bacteria.

Regarding the Fec system, an interesting second level of regulation has recently linked iron uptake to carbon metabolism. Bioinformatic methods were used to explore more than 400 microarray data sets of *E. coli* to find new regulatory interactions (Faith et al. 2007). PdhR, a pyruvate sensing repressor, was shown to regulate FecA expression. Both the activated FecI and the activated PdhR are necessary for maximal expression of the *fec* operon.

2.15.3.3 AraC- and LysR-Type Regulators

The first $Fe^{2+}Fur$ regulated AraC-like regulators were identified in pyochelin, yersiniabactin, and alcaligin biosynthesis and transport. In all of these cases, the corresponding siderophores act as coinducers in the system as it is known for many AraC-like regulators. With a catechol as a coinducer, DhbR acts as an activator in catechol biosynthesis of *Brucella abortus* (Anderson et al. 2008). The uptake of the xenosiderophore ferrioxamine B is induced in *Vibrio furnissii* via the AraC-like regulator DesR (Tanabe et al. 2011).

An AraC-like regulator, MpeR, was found in pathogenic *Neisseria gonorrhoeae* to stimulate expression of the FetA enterobactin outer membrane transporter (Hollander et al. 2011) under iron-limiting conditions. Interestingly, MpeR had been found previously to act as a repressor of the Mtr transport system under iron-limiting conditions. The Mtr transport system exports structurally divergent hydrophobic antibiotics and detergents. It was not possible to identify a coinducer like those found with most other AraC-like regulators.

LysR-type regulators belong to the most common prokaryotic regulators which control a functionally very diverse set of genes. The OxyR proteins regulating the response to oxidative stress belong to the LysR protein family. They act in line with their oxidative stress response and also on iron metabolism, since Fe^{2+} reacting with oxygen is a major source of toxic radicals (Faulkner and Helmann 2011). In addition to the induction of repair enzymes, the distribution of metal ions in the cell is changed through OxyR. A major aim for the cells is to lower the ferrous iron concentration in the cytosol. Therefore, it is not surprising that there are also LysR-type regulators which control specific iron uptake genes. An example of this is the LysR-type activator HmuB from *Vibrio mimicus*, which stimulates the expression of heme uptake genes (Tanabe et al. 2010).

2.16 Concluding Remarks

Iron is an essential element, since it is contained in many redox enzymes. Although available in large quantities on earth, its insolubility in the Fe^{3+} state makes it difficult for all aerobic organisms to acquire enough iron for their cell metabolism. To cope with this problem, all organisms invented intricate iron transport systems. Since microbes live under very diverse environmental conditions, they had to develop a large variety of iron uptake mechanisms to get enough iron into the cells. Competition for iron is a means of eukaryotic cells to keep microbial invaders under control. Microbes on the other hand developed systems to overcome iron limitation imposed by eukaryotes. Moreover, they had to find ways to control the intracellular iron concentrations to avoid formation of damaging oxygen radicals by a surplus of iron. The control mechanisms go beyond maintenance of iron homeostasis and affect pathways that are directly or indirectly related to iron metabolism. The iron regulatory network is one of the central control mechanisms of cell metabolism.

References

Adhikari P, Kirby SD, Nowalk AJ et al (1995) Biochemical characterization of a *Haemophilus influenzae* periplasmic iron transport operon. J Biol Chem 270:25142–25149

Africa LA, Murphy ER, Egan NR et al (2011) The iron-responsive Fur/RyhB regulatory cascade modulates the *Shigella* outer membrane protease IcsP. Infect Immun 79:4543–4549

Ahn BE, Cha J, Lee EJ, Han AR et al (2006) Nur, a nickel-responsive regulator of the Fur family, regulates superoxide dismutases and nickel transport in *Streptomyces coelicolor*. Mol Microbiol 59:1848–1858

Alt S, Burkard N, Kulik A et al (2011) An artificial pathway to 3,4 dihydroxybenzoic acid allows generation of new aminocoumarin antibiotic recognized by catechol transporters of *E. coli*. Chem Biol 18:304–313

An YJ, Ahn BE, Han AR, Kim HM et al (2009) Structural basis for the specialization of Nur, a nickel-specific Fur homolog, in metal sensing and DNA recognition. Nucleic Acids Res 37:3442–3451

Anderson ES, Paulley JT, Roop RM (2008) The AraC-like transcriptional regulator DhbR is required for maximum expression of the 2,3-dihydroxybenzoic acid biosynthesis genes in *Brucella abortus* 2308 in response to iron deprivation. J Bacteriol 190:1838–1842

Angerer A, Gaisser S, Braun V (1990) Nucleotide sequence of the *sfuA*, *sfuB* and *sfuC* genes of *Serratia marcescens* suggest a periplasmic–binding-protein–dependent iron transport mechanism. J Bacteriol 172(572):578

Angerer A, Klupp B, Braun V (1992) Iron transport systems of *Serratia marcescens*. J Bacteriol 174:1378–1387

Aranda J, Cortes P, Garrido ME et al (2009) Contribution of the FeoB transporter to *Strepococcus suis* virulence. Int Microbiol 12:137–143

Argaman L, Elgrably-Weiss M, Hershko T et al (2012) RelA protein stimulates the activity of RyhB small RNA by acting on RNA-binding protein Hfq. Proc Natl Acad Sci USA 109: 4621–4626

Beasley FC, Marolda CL, Cheung J et al (2011) *Staphylococcus aureus* transporters Hts, Sir, and Sst capture iron liberated from human transferrin by staphyloferrin A, staphyloferrin B and catecholamine stress hormones, respectively, and contribute to virulence. Infect Immun 79:2345–2355

Bister B, Bischoff D, Nicholson GJ et al (2004) The structure of salmochelins: C-glucosylated enterobactins of *Salmonella enterica*. Biometals 17:471–481

Bjursell MK, Martens EC, Gordon JI (2006) Functional genomic and metabolic studies of the adaptation of a prominent adult human gut symbiont, *Bacteroides thetaiotaomicron*, to the suckling period. J Biol Chem 281:36269–36279

Blanvillain S, Meyer D, Boulanger A et al (2007) Plant carbohydrate scavenging through *tonB*-dependent receptors: a feature shared by phytopathogenic and aquatic bacteria. PLoS ONE 2:e224

Bleuel C, Grosse C, Taudte N, Scherer J, Wesenberg D, Krauss GJ, Nies DH, Grass G (2005) TolC is involved in enterobactin efflux across the outer membrane of *Escherichia coli*. J Bacteriol 187:6701–6707

Bosello M, Robbel L, Linne U et al (2011) Biosynthesis of the siderophore rhodochelin requires the coordinated expression of three independent gene clusters in *Rhodococcus jostii* RhA1. J Amer Chem Soc 133:4587–4595

Braun V (1999) Active transport of siderophore-mimicking antibacterials across the outer membrane. Drug Resist Updat 2:363–369

Braun V (2010) Outer membrane signaling in Gram-negative bacteria. In: Krämer R, Jung K (eds) Bacterial Signaling. Wiley-Blackwell, Weinheim, pp 117–133

Braun M, Endriss F, Killmann H et al (2003a) In vivo reconstitution of the FhuA transport protein of *Escherichia coli* K-12. J Bacteriol 185:5508–5518

Braun V, Gaisser S, Herrmann C et al (1996) Energy-coupled transport across the outer membrane of *Escherichia coli*: ExbB binds ExbD and TonB in vitro, and Leucine 132 in the periplasmic region and aspartate 25 in the transmembaren region are important for ExbB activity. J Bacteriol 178:2836–2845

Braun V, Günthner K, Hantke K et al (1983) Intracellular activation of albomycin in *Escherichia coli* and *Salmonella typhimurium*. J Bacteriol 156:308–315

Braun V, Hantke K, Köster W (1998) Bacterial iron transport: mechanisms, genetics, and regulation. In: Sigel A, Sigel H (eds) Metal ions in biological systems. Marcel Dekker, New York

Braun V, Herrmann C (2004a) Evolutionary relationship of uptake systems for biopolymers in *Escherichia coli*: cross-complementation between the TonB-ExbB-ExbD and the TolA-TolQ-TolR proteins. Mol Microbiol 8:261–268

Braun V, Herrmann C (2004b) Point mutations in transmembrane helices 2 and 3 of ExbB and TolQ affect their activities in *Escherichia coli* K-12. J Bacteriol 186:4402–4406

Braun M, Killmann K, Maier E et al (2002a) Diffusion through channel derivatives of the *Escherichia coli* FhuA transport protein. Eur J Biochem 269:4948–4959

Braun V, Mahren S (2005) Transmembrane transcriptional control (surface signaling) of the *Escherichia coli* Fec type. FEMS Microbiol Rev 29:673–684

Braun V, Mahren S (2007) Transfer of energy and information across the periplasm in iron transport and regulation. In: Ehrmann M (ed) The periplasm. ASM Press, Washington, DC, pp 276–286

Braun V, Mahren S, Ogierman M (2003b) Regulation of the FecI type ECF sigma factor by transmembrane signaling. Curr Opin Microbiol 6:173–180

Braun V, Mahren S, Sauter A (2006) Gene regulation by transmembrane signaling. Biometals 18:507–517

Braun V, Patzer SI, Hantke K (2002b) TonB-dependent colicins and microcins: modular design and evolution. Biochimie 84:365–380

Braun V, Pramanik A, Gwinner T et al (2009) Sideromycins: tools and antibiotics. Biometals 22:3–13

Brett PJ, Burtnick MN, Fenno JC et al (2008) *Treponema denticola* TroR is a manganese- and iron-dependent transcriptional repressor. Mol Microbiol 70:396–409

Carvalho CCCR, Marques MPC, Fernandes P (2011) Recent achievements on siderophore production and application. Recent Pat Biotechnol 5:183–198

Cartron, ML Maddocks S, Gillingham P et al (2006) Feo–transport of ferrous iron into bacteria. Biometals 10:143–157

Chatfield CH, Mulhern BJ, Burnside NP et al (2011) *Legionella pneumophila* LbtU acts as a novel, TonB-independent receptor for the legiobactin siderophore. J Bacteriol 194:1563–1575

Chu BC, Garcia Herrero A, Johnson TH et al (2010) Siderophore uptake in bacteria and the battle for iron with the host; a bird's eye view. Biometals 23:601–611

Chu BCH, Peacock RS, Vogel HJ (2007) Bioinformatic analysis of the TonB family. Biometals 20:467–483

Chu BC, Vogel HJ (2011) A structural and functional analysis of type III periplasmic and substrate binding proteins: their role in bacterial siderophore and heme uptake. Biol Chem 392:39–52

Clarke TE, Ku SY, Vogel H et al (2000) The structure of the ferric siderophore binding protein FhuD complexed with gallichrome. Nat Struct Biol 7:287–291

Clarke TE, Braun V, Winkelmann G et al (2002) X-ray crystallographic structures of the *Escherichia coli* periplasmic protein FhuD bound to hydroxamate-type siderophores and the antibiotic albomycin. J Biol Chem 277:13966–13972

Cornelis P, Andrews SC (2010) Iron uptake and homeostasis in microorganisms. Caister Academic Press, Norfolk

Correnti C, Strong RK (2012) Mammalian siderophores, siderophores-binding lipocalins, and the labile iron pool. J Biol Chem 287:13524–13531

da Silva Neto JF, Braz VS, Italiani VC et al (2009) Fur controls iron homeostasis and oxidative stress defense in the oligotrophic alpha-proteobacterium *Caulobacter crescentus*. Nucleic Acids Res 37:4812–4825

Dian C, Vitale S, Leonard GA et al (2011) The structure of the *Helicobacter pylori* ferric uptake regulator Fur reveals three functional metal binding sites. Mol Microbiol 79:1260–1275

Diaz-Mireles E, Wexler M, Sawers G et al (2004) The Fur-like protein Mur of *Rhizobium leguminosarum* is a Mn(2 +)-responsive transcriptional regulator. Microbiology 150: 1447–1456

Draper RC, Martin LW, Beare PA et al (2011) Differential proteolysis of sigma regulators controls cell-surface signalling in *Pseudomonas aeruginosa*. Mol Microbiol 82:1444–1453

Eisenbeis S, Lohmiller S, Valdebenito M et al (2008) NagA-dependent uptake of N-acetyl-glucosamine and N-acetyl-chitin oligosaccharides across the OM of *Caulobacter crescentus*. J Bacteriol 190:5230–5238

Eng ET, Jalilian AR, Spasov KA et al (2008) Characterization of a novel prokaryotic GDP dissociation inhibitor domain from the G protein coupled membrane protein FeoB. J Mol Biol 375:1086–1097

Faith JJ, Hayete B, Thaden JT et al (2007) Large-scale mapping and validation of *Escherichia coli* transcriptional regulation from a compendium of expression profiles. PLoS Biol 5:e8

Faulkner MJ, Helmann JD (2011) Peroxide stress elicits adaptive changes in bacterial metal ion homeostasis. Antioxid Redox Signal 15:175–189

Faulkner MJ, Ma Z, Fuangthong M et al (2012) Derepression of the *Bacillus subtilis* PerR peroxide stress response leads to iron deficiency. J Bacteriol 194:1226–1235

Ferguson AD, Braun V, Fiedler HP et al (2000) Crystal structure of the antibiotic albomycin in complex with the OM transporter FhuA. Prot Sci 9:956–963

Ferguson AD, Koding J, Walker G et al (2001) Active transport of an antibiotic rifamycin derivative by the outer membrane protein FhuA. Structure 9:707–716

Ferguson AD, Amezcua CA, Halabi NM et al (2007) Signal transduction pathway of TonB-dependent transporters. Proc Natl Acad ScI USA 104:513–518

Ferguson AD, Hofmann EE, Coulton JEW et al (1998) Siderophore-mediated iron transport: crystal structure of FhuA with bound lipopolysaccharide. Science 282:2215–2220

Ferguson AD, Chakraborty R, Smith BS et al (2002) Structural basis of gating by the OM transporter FecA. Science 295:1715–1719

Fischbach MA, Smith KD, Sato S et al (2006) The pathogen-associated *iroA* gene cluster mediates bacterial evasion of lipocalin 2. Proc Natl Acad Sci USA 103:16502–16507

Fisher CR, Davies NMLL, Wyckoff EE et al (2009) Genetics and virulence association of the *Shigella flexneri* Sit iron transport system. Infect Immun 77:1992–1999

Fleischhacker AS, Kiley PJ (2011) Iron-containing transcription factors and their roles as sensors. Curr Opin Chem Biol 15:335–341

Floh TH, Smith KD, Sato S et al (2004) Lipocalin 2 mediates an innate immune response to bacterial infection by sequestrating iron. Nature 4232:917–921

Fournier C, Smith A, Delepelaire P (2011) Haem release from haemopexin by HuxA allows *Haemophilus influenzae* to escape host nutritional immunity. Mol Microbiol 80:133–148

Furrer JL, Sanders DN, Hook-Barnard IG et al (2002) Export of the siderophore enterobactin in *Escherichia coli*: involvement of a 43 kDa membrane exporter. Mol Microbiol 44:1225–1234

Ganz T (2006) Hepcidin-a peptide hormone at the interface of innate immunity and iron metabolism. Curr Top Microbiol Immunol 306:183–198

Gauglitz JM, Zhou H, Butler A (2012) A suite of citrate-derived siderophores from a marine *Vibrio* species isolated following the Deepwater Horizon oil spill. J Inorg Chem 107:90–95

Gonzalez A, Bes MT, Peleato ML et al (2011) Unravelling the regulatory function of FurA in *Anabaena* sp. PCC 7120 through 2-D DIGE proteomic analysis. J Proteomics 74:660–671

Greenwald J, Hoegy F, Nader M et al (2007) Real time fluorescence resonance transfer visualization of ferric pyoverdine uptake in *Pseudomonas aeruginosa*. The role of ferrous iron. J Biol Chem 282:2987–2995

Greenwood KT, Luke RKJ (1978) Enzymatic hydrolysis of enterochelin and its iron complex in *Escherichia coli*. Properties of enterochelin esterase. Biochim Biophys Acta 525:209–218

Grinter R, Milner J, Walker D (2012) Ferredoxin containing bacteriocins suggest a novel mechanisms of iron uptake in *Pectobacterium* spp. PLoS ONE 7(1–9):e33033

Groeger W, Köster W (1998) Transmembrane topology of the two FhuB domains representing the hydrophobic components of bacterial ABC transporters involved in uptake of siderophores, haem and vitamin B_{12}. Microbiology 144:2759–2769

Grosse C, Scherer J, Koch D et al (2006) A new ferrous iron-uptake transporter, EfeU (YcdN) from *Escherichia coli*. Mol Microbiol 62:120–131

Guilfoyle A, Maher MJ, Rapp M et al (2009) Structural basis of GDP release and gating in G protein coupled Fe^{2+} transport. EMBO J 28:2677–2685

Gumbart J, Wiener MC, Tajkhorshid E (2007) Mechanism of force propagation in TonB-dependent OM transport. Biophys J 93:496–504

Günter-Seeboth K, Schupp T (1995) Cloning and sequence analysis of the *Corynebacterium diphtheriae* dtxR homologue from *Streptomyces lividans* and *S. pilosus* encoding a putative iron repressor protein. Gene 166:117–119

Hammer ND, Scaar EP (2011) Molecular mechanism of *Staphylococcus aureus* iron acquisition. Annu Rev Microbiol 65:129–147

Hannauer M, Yeterian E, Martin LW et al (2010a) An efflux pump is involved in secretion of newly synthesized siderophore by *Pseudomonas aeruginosa*. FEBS Lett 584:4451–4455

Hannauer M, Barda Y, Mislin GLA et al (2010b) The ferrichrome uptake pathway in *Pseudomonas aeruginosa* involves an iron release mechanism with acylation of the siderophore and recycling of the modified desferriferrichrome. J Bacteriol 192:1212–1220

Hannauer M, Schäfer M, Hoegy F et al (2012) Biosynthesis of the pyoverdine siderophore of *Pseudomonas aeruginosa* involves precursors with myristic and myristoleic acid chain. FEBS Lett 586:96–101

Hantke K (1987) Ferrous iron transport mutants in *Escherichia coli* K-12. FEMS Microbiol Lett 44:53–57

Hartmann A, Braun V (1980) Iron transport in *Escherichia coli*: uptake and modification of ferrichrome. J Bacteriol 143:246–255

Hattori M, Jin Y, Nishimasu H et al (2009) Structural basis of novel interactions between the small-GTPase and the GDI-like domains in prokaryotic FeoB iron transporter. Structure 17:1345–1355

Helbig S, Braun V (2011) Mapping functional domains of colicin M. J Bacteriol 193:815–821

Hider RC, Kong X (2010) Chemistry and biology of siderophores. Nat Prod Rep 27:637–657

Higgs PI, Larsen RA, Postle K (2002) Quantification of known components of the *Escherichia coli* TonB energy transducing system: TonB, ExbB, ExbD, and FepA. Mol Microbiol 44:271–281

Hollander A, Mercante AD, Shafer WM et al (2011) The iron-repressed, AraC-like regulator MpeR activates expression of *fetA* in *Neisseria gonorrhoeae*. Infect Immun 79:4764–4776

Hopkinson BM, Barbeau KA (2011) Iron transporters in marine prokaryotic genomes and metagenomes. Environ Microbiol 14:114–128

Ji C, Juarez-Hernandez RE, Miller MJ (2012) Exploiting bacterial iron acquisition: siderophore conjugates. Future Med Chem 4:297–313

Johnston AW, Todd JD, Curson AR et al (2007) Living without Fur: the subtlety and complexity of iron-responsive gene regulation in the symbiotic bacterium *Rhizobium* and other alpha-proteobacteria. Biometals 20:501–511

Joseph B, Jeschke G, Goetz BA et al (2011) Transmembrane gate movements in the type II ATP-binding cassette (ABC) importer BtuCD-F during nucleotide cycle. J Biol Chem 286:41008–41017

Kehres DG, Janakiraman A, Slauch JM et al (2002) SitABCD is the alkaline Mn^{2+} transporter of *Salmonella enterica* serovar Typhimurium. J Bacteriol 184:3159–3166

Killmann H, Herrmann C, Wolff H et al (1998) Identification of a new site for ferrichrome transport by comparison of the FhuA proteins of *Escherichia coli, Salmonella paratyphi B, Salmonella typhimurium*, and *Pantoea agglomerans*. J Bacteriol 180:3845–3852

Koebnik R (2005) TonB-dependent trans-envelope signaling: the exception of the rule? Trends Microbiol 13:343–347

Köster W (2005) Cytoplasmic membrane iron permease systems in the bacterial cell envelope. Frontiers Biosci 10.462–477

Köster S, Wehner M, Herrmann C et al (2009) Structure and function of the FeoB G-domain from *Methanococcus jannaschii*. J Mol Biol 392:405–419

Krewulak KD, Shepherd CM, Vogel HJ (2005) Molecular dynamics simulations of the periplasmic ferric-hydroxamate binding protein FhuD. Biometals 18:375–386

Krieg S, Huché F, Diederichs K, Izadi-Pruneyre N et al (2009) Heme uptake across the OM as revealed by crystal structures of the receptor-heme complex. Proc Natl Acad Sci USA 106:1045–1050

Kustusch RJ, Kuehl C, Crosa JH (2011) Power plays: iron transport and energy transduction in pathogenic vibrios. Biometals 24:559–566

Létoffé S, Ghigo JM, Wandersman C (1994) Secretion of the *Serratia marcescens* HasA protein by an ABC transporter. J Bacteriol 176:5327–5377

Lewinson O, Lee AT, Locher KP, Rees DC (2010) A distinct mechanism for the ABC transporter BtuCD-BtuF revealed by the dynamics of complex formation. Nat Struct Biol 17:332–339

Li N, Zhang C, Li B et al (2012) An unique iron coordination in the iron-chelating molecule vibriobactin helps *Vibrio cholerae* evade the mammalian siderocalin-mediated immune response. J Biol Chem 287:8912–8919

Locher KP, Lee AT, Rees DC (2002) The *E. coli* BtucD structure: a framework for ABC transporter architecture and mechanism. Science 296:1091–1098

Locher KP, Rees B, Koebnik R et al (1998) Transmembrane signaling across the ligand-gated FhuA receptor: crystal structures of free and ferrichrome-bound states reveal allosteric changes. Cell 95:771–778

Lohmiller S, Hantke K, Patzer SI et al (2008) TonB-dependent maltose transport by *Caulobacter crescentus*. Microbiology 154:1748–1754

Lo M, Murray GL, Khoo CA et al (2010) Transcriptional response of *Leptospira interrogans* to iron limitation and characterization of a PerR homolog. Infect Immun 78:4850–4859

Louvel H, Bommezzadri S, Zidane N et al (2006) Comparative and functional genomic analyses of iron transport and regulation in *Leptospira* spp. J Bacteriol 188:7893–7904

Louvel H, Kanai T, Atomi H et al (2009) The Fur iron regulator-like protein is cryptic in the hyperthermophilic archaeon *Thermococcus kodakaraensis*. FEMS Microbiol Lett 295:117–128

Lukacik P, Barnard TJ, Keller PW et al (2012) Structural engineering of a phage lysin that targets Gram-negative pathogens. Proc Natl Acad Sci USA 109:9857–9862

Marlovits TC, Haase W, Herrmann C et al (2003) The membrane protein FeoB contains an intramolecular G protein essential for Fe(II) uptake in bacteria. Proc Natl Acad Sci USA 99:16243–16248

Matzanke BF, Anemüller S, Schünemann V et al (2004) FhuF, part of a siderophore-reductase system. Biochemistry 43:1386–1392

Mazmanian SK, Skaar EP, Gaspar AH et al (2003) Passage of heme –iron across the envelope of *Staphylococcus aureus*. Science 299:906–909

Mey AR, Wyckoff EE, Hoover LA et al (2008) *Vibrio cholerae* VciB promotes iron uptake via ferrous iron transporters. J Bacteriol 190:5953–5962

Miethke M, Hou J, Marahiel MA (2011a) The siderophore interacting protein YqjH acts as a ferric reductase in different iron assimilation pathways of *Escherichia coli*. Biochemistry 50:10951–10964

Miethke M, Pierik AJ, Peuckert F et al (2011b) Identification and characterization of a nove-type ferric siderophore reductase from a gram-positive extremophile. J Biol Chem 286:2245–2260

Moraes TF, Yu RH, Strynadka NC et al (2009) Insights into the bacterial transferrin receptor: the structure of transferrin-binding protein B from *Actinobacillus pleuropneumoniae*. Mol Cell 35:523–533

Müller SI, Valdebenito M, Hantke K (2009) Salmochelin, the long-overlooked catecholate siderophore of Salmonella. Biometals 22:691–695

Neugebauer H, Herrmann C, Kammer W et al (2005) ExbBD-dependent transport of maltodextrins through the novel MalA protein across the outer membrane of *Caulobacter cresescentus*. J Bacteriol 187:8300–8311

Newton SM, Trinh V, Pi H et al (2010) Direct measurements of the OM stage of ferric enterobactin transport postuptake binding. J Biol Chem 285:17488–17497

Nicolaisen K, Hahn A, Valdebenito M et al (2010) The interplay between siderophore secretion and coupled iron and copper transport in the heterocyst-forming cyanobacterium *Anabaena* sp.PCC 7120. Biochim Biophys Acta 1798:2131–2140

Noinaj N, Easley NC, Oke M et al (2012) Structural basis for iron piracy by pathogenic *Neisseria*. Nature 483:53–58

Noinaj N, Guillier M, Barnard TJ et al (2010) TonB dependent transporters: regulation, structure and function. Annu Rev Microbiol 64:43–60

Oldham ML, Chen J (2011) Crystal structure of the maltose transporter in a pretranslocation intermediate state. Science 332:1201–1205

Ollinger J, Song KB, Antelmann H et al (2006) Role of the Fur regulon in iron transport in *Bacillus subtilis*. J Bacteriol 188:3664–3673

Ollis AA, Postle K (2012) ExbD mutants define initial stages in TonB energization. J Mol Biol 415:237–247

Patzer SI, Braun V (2010) Gene cluster involved in the biosynthesis of griseobactin, a catechol-peptide siderophore of *Streptomyces* sp ATCC 700974. J Bacteriol 192:426–435

Patzer SI, Albrecht R, Braun V, et al (2012) Structure and mechanistic studies of pesticin, a bacterial homolog of phage lysozymes. J Biol Chem 287(28):23381–23396

Patzer SI, Hantke K (1998) The ZnuABC high-affinity zinc uptake system and its regulator Zur in *Escherichia coli*. Mol Microbiol 28:1199–1210

Patzer SI, Hantke K (2001) Dual repression by Fe(2 +)-Fur and Mn(2 +)-MntR of the *mntH* gene, encoding an NRAMP-like Mn(2 +) transporter in *Escherichia coli*. J Bacteriol 183:4806–4813

Pawelek PD, Croteau N, Ng-Tow-Hing C et al (2006) Structure of TonB in complex with FhuA *E. coli* OM receptor. Science 312:1399–1402

Pohl E, Haller JC, Mijovilovich A et al (2003) Architecture of a protein central to iron homeostasis: crystal structure and spectroscopic analysis of the ferric uptake regulator. Mol Microbiol 47:903–915

Postle K, Larsen RA (2007) TonB-dependent energy transduction between outer and cytoplasmic membranes. Biometals 20:453–465

Pramanik A, Braun V (2006) Albomycin uptake via a ferric hydroxamate transport system of *Streptococcus pneumoniae* R6. J Bacteriol 188:3878–3886

Pramanik A, Stroeher UW, Krejci J et al (2007) Albomycin is an effective antibiotic, as exemplified with *Yersinia enterocolitica* and *Streptococcus pneumoniae*. Int J Med Micriobiol 297:459–469

Pramanik A, Hauf W, Hoffmann J et al (2011) Oligomeric structure of ExbB and ExbB-ExbD isolated from *Escherichia coli* as revealed by LILBID mass spectrometry. Biochemistry 50:8950–8956

Pramanik A, Zhang F, Schwarz H et al (2010) ExbB protein in the cytoplasmic membrane of *Escherichia coli* forms a stable oligomer. Biochemistry 49:8721–8728

Prevost K, Salvail H, Desnoyers G et al (2007) The small RNA RyhB activates the translation of shiA mRNA encoding a permease of shikimate, a compound involved in siderophore synthesis. Mol Microbiol 64:1260–1273

Pugsley AP, Zimmermann W, Wehrli W (1987) High efficiency uptake of a rifamycin derivative via the FhuA-TonB-dependent uptake route in *Escherichia coli*. J Gen Microbiol 133: 3505–3511

Reimmann C (2012) Inner-membrane transporters for the siderophores pyochelin in *Pseudomonas aeruginosa* and enantio-pyochelin in *Pseudomonas fluorescens* display different enantioselectivities. Microbiology 158:1317–1324

Saken E, Rakin A, Heesemann J (2000) Molecular characterization of a novel siderophore-independent iron transport system in Yersinia. Int J Med Microbiol 290:51–60

Salvail H, Lanthier-Bourbonnais P, Sobota JM et al (2010) A small RNA promotes siderophore production through transcriptional and metabolic remodeling. Proc Natl Acad Sci USA 107:15223–15228

Schauer K, Gouget B, Carrière M et al (2007) Novel nickel transport mechanism across the bacterial OM energized by the TonB/ExbB/ExbD machinery. Mol Microbiol 63:1054–1068

Schauer K, Rodionov DA, de Reuse H (2008) New substrates for TonB-dependent transport: do we only see the "tip of the iceberg"? Trends Biochem Sci 33:330–338

Schmid AK, Pan M, Sharma K et al (2011) Two transcription factors are necessary for iron homeostasis in a salt-dwelling archaeon. Nucleic Acids Res 39:2519–2533

Seipke RF, Song L, Bicz J et al (2011) The plant pathogen *Streptomyces scabies* 87–22 has a functional pyochelin biosynthetic pathway that is regulated by TetR- and AfsR-family proteins. Microbiology 157:2681–2693

Shin JH, Oh SY, Kim SJ et al (2007) The zinc-responsive regulator Zur controls a zinc uptake system and some ribosomal proteins in *Streptomyces coelicolor*A3(2). J Bacteriol 189:4070–4077

Shultis DD, Purdy MD, Branchs CN et al (2006) Outer membrane active transport: structure of the BtuB:TonB complex. Science 312:1396–1399

Singh PK, Parsek PR, Greenberg EP et al (2002) A component of innate immunity prevents bacterial biofilm development. Nature 417:552–555

Small SK, Puri S, O'Brian MR (2009) Heme-dependent metalloregulation by the iron response regulator (Irr) protein in *Rhizobium* and other Alpha-proteobacteria. Biometals 22:89–97

Stefanska AL, Fulston M, Houge-Frydrych CSV et al (2000) A potent seryl tRNA synthetase inhibitor SB-217452 isolated from a *Streptomyces* species. J Antibiot 53:1346–1353

Stojiljkovic I, Cobeljic M, Hantke K (1993) *Escherichia coli* K-12 ferrous iron uptake mutants are impaired in their ability to colonize the mouse intestine. FEMS Microbiol Lett 108: 111–115

Stork M, Bos MP, Jongerius I et al (2010) An OM receptor of *Neisseria meningitidis* involved in zinc acquisition with vaccine potential. PLoS Pathog 7:e100969

Strange HR, Zola TA, Cornelissen CN (2011) The *fbpABC* operon is required for TonB-independent utilization of xenosiderophores *by Neisseria gonorrhoeae* strain FA19. Infect Immun 79:267–278

Su YC, Chin KH, Hung HC et al (2010) Structure of *Stenotrophomonas maltophila* FeoA complexed with zink: unique prokaryotic SH3-domain protein that possibly acts a bacterial ferrous iron-transport activating factor. Acta Crystallogr Sect F 66:636–642

Swayne C, Postle K (2011) Taking the *Escherichia coli* TonB transmembrane domain "offline"? Nonprotonatable Asn substitutes fully for TonB His20. J Bacteriol 193:6393–6701

Tanabe T, Funahashi T, Miyamoto K et al (2011) Identification of genes, *desR* and *desA*, required for utilization of desferrioxamine B as a xenosiderophore in *Vibrio furnissii*. Biol Pharm Bull 34:570–574

Tanabe T, Funahashi T, Moon YH et al (2010) Identification and characterization of a *Vibrio mimicus* gene encoding the heme/hemoglobin receptor. Microbiol Immunol 54:606–617

Udho E, Jakes KS, Buchanan SK et al (2009) Reconstitution of bacterial OM TonB-dependent transporters in planar lipid bilayer membranes. Proc Natl Acad Sci USA 106:21990–21995

Valdebenito M, Müller SI, Hantke K (2007) Special conditions allow binding of the siderophore salmochelin to siderocalin (NGAL-lipocalin). FEMS Microbiol Lett 277:182–187

Velayudhan J, Hughes NJ, McColm AA et al (2000) Iron acquisition and virulence in *Helicobacter pylori*: a major role for FeoB, a high affinity ferrous iron transporter. Mol Microbiol 37:274–286

Vinella D, Albrecht C, Cashel M et al (2005) Iron limitation induces SpoT-dependent accumulation of ppGpp in *Escherichia coli*. Mol Microbiol 56:958–970

Vraspir JM, Butler A (2009) Chemistry of marine ligands and siderophores. Annu Rev Mar Sci 1:43–63

Wandersman C (2010) Haem uptake and iron extraction by bacteria. In: Cornelis P, Andrews SC (eds) Iron uptake and homeostasis in microorganisms. Claister Academic Pres, Norfolk

Wandersman C, Stojiljkovic I (2000) Bacterial heme sources: the role of heme, heme protein receptors and hemophores. Curr Op Microbiol 3:215–220

Wennerhold J, Bott M (2006) The DtxR regulon of *Corynebacterium glutamicum*. J Bacteriol 188:2907–2918

Wyckoff EE, Payne SM (2011) The *Vibrio cholerae* VctPDGC system transports catechol siderophores and a siderophore-free iron ligand. Mol Microbiol 81:1556–1458

Xiao Q, Jiang X, Moore KJ et al (2011) Sortase independent and dependent systems for acquisition of haem and haemoglobin in *Listeria monocytogenes*. Mol Microbiol 80:1581–1597

Yue WW, Grizot S, Buchanan SK (2003) Structural evidence for iron-free citrate and ferric citrate binding to the TonB-dependent OM transporter FecA. J Mol Biol 332:353–368

Zimmermann L, Angerer A, Braun V (1989) Mechanistically novel iron(III) transport system in *Serratia marcescens*. J Bacteriol 171:238–243

Chapter 3
Iron Transport Systems and Iron Homeostasis in *Pseudomonas*

Pierre Cornelis

Abstract During the last few years the knowledge about iron uptake and homeostasis in *Pseudomonas* has increased enormously. These very versatile bacteria can adapt to widely different ecological niches. It is therefore not surprising that *Pseudomonas* has a remarkable ability to take up iron and balance iron levels in the cell. The fluorescent pseudomonads, the best known species being *P. aeruginosa*, *P. putida*, *P. syringae*, and *P. fluorescens*, all produce a fluorescent pigment called pyoverdine, which serves as the major siderophore to capture iron (III). Pyoverdines are complex peptidic structures and each species produces its own pyoverdine siderophore and the corresponding receptor at the level of the outer membrane, meaning that both receptors and pyoverdines co-evolved. A peculiarity of the pyoverdine-mediated iron uptake is the release of iron since it takes place in the periplasm. Many pseudomonads produce a second siderophore of lesser affinity as well, such as pyochelin, enantio-pyochelin, pseudomonin, yersiniabactin, thioquinolobactin, achromobactin, and PDTC, which have other functions next to their role in iron uptake, such as antimicrobial and catalytic activity. A remarkable characteristic of fluorescent pseudomonads is their capacity to scavenge siderophores produced by other microorganisms (xenosiderophores) via a plethora of different outer membrane receptors. Heme is another source of iron that can be used by pseudomonads, animal pathogens such as *P. aeruginosa* and *P. entomophila*, having three different heme uptake systems. Finally, some pseudomonads have the capacity to take up iron (II). The regulation of iron homeostasis in fluorescent pseudomonads is quite elaborate and multi-layered, involving the master regulator Fur and secondary regulators, including sigma factors, two-component systems regulators, and small RNAs. Finally, we will

P. Cornelis (✉)
Department of Bioengineering Sciences, Research Group Microbiology, Vrije Universiteit Brussel and VIB Structural Biology, Pleinlaan 2 1050 Brussels, Belgium
e-mail: pcornel@vub.ac.be

R. Chakraborty et al. (eds.), *Iron Uptake in Bacteria with Emphasis on E. coli and Pseudomonas*, SpringerBriefs in Biometals, DOI: 10.1007/978-94-007-6088-2_3, © The Author(s) 2013

present evidence that there is cross-talk between the quorum sensing regulon and iron homeostasis as well as between the response to oxidative stress and the control of iron uptake.

Keywords *Pseudomonas* · Iron · Siderophores · Pyoverdines · Secondary siderophores · Xenosiderophores · TonB-dependent receptors · Heme uptake · Fur · ECF-sigmas · Regulators · Oxidative stress

3.1 The Fluorescent Pseudomonads

Pseudomonads belong to the γ-proteobacteria and are motile, having generally a single polar flagellum. These bacteria are metabolically highly versatile and consequently they are able to colonize different habitats (water, the rhizosphere, plant leaves, insects, mammals) (Goldberg 2000). *Pseudomonas putida* is known for its ability to degrade aromatic compounds, including xenobiotics (Nelson et al. 2002; Weinel et al. 2002). *Pseudomonas fluorescens* is a soil bacterium often found in association with plant roots where it exerts a beneficial effect thanks to the production of secondary metabolites active against phytopathogens (Paulsen et al. 2005). Some fluorescent pseudomonads are pathogens for plants (*P. syringae*), insects (*P. entomophila*), or humans (*P. aeruginosa*) (Stover et al. 2000; Buell et al. 2003; Joardar et al. 2005; Vodovar et al. 2006). The human opportunistic pathogen *P. aeruginosa* is also able to infect other hosts, such as amoebas, nematodes, insects, and plants (Mahajan-Miklos et al. 1999; Pukatzki et al. 2002; Walker et al. 2004; Abd et al. 2008; Apidianakis and Rahme 2009). Due to their adaptability and highly versatile metabolism, it is not surprising that fluorescent pseudomonads have large genomes (from 6 to 7 Mb), among the largest among Gram-negative bacteria.

3.2 Siderophore-Mediated Iron Uptake in Gram-Negative Bacteria

For almost all microorganisms iron is an essential element that, although abundant in nature, is not easily available under aerobic conditions, due to the poor solubility of the oxidized Fe^{3+} form (Andrews et al. 2003). Iron is essential due to its involvement in important metabolic processes including respiration ([Fe–S]-containing ferredoxins, heme-containing cytochromes) and key enzymatic reactions (Andrews et al. 2003). In the presence of oxygen at physiological pH iron is oxidized (Fe^{3+}) and is highly insoluble due to the formation of iron hydroxides. In order to solubilize iron, bacteria produce strong extracellular Fe^{3+} chelators, termed siderophores (Braun and Killmann 1999). Gram-negative bacteria, including pseudomonads, have in the outer membrane TonB-dependent outer membrane receptors, which bind the iron-loaded siderophore complex. These receptors are large gated porins with 22 β-strands forming a β-barrel. The pore is constricted by the N-terminal domain of the protein (the "cork"). The N-terminal end of the receptor facing the periplasm also contains a domain termed the TonB-box, which interacts with the inner membrane protein TonB. This inner membrane protein, together with ExbB and ExbD relays the energy of the proton motive force in order to open the receptor gate, resulting in the transport of the ferric complex into the periplasm where it is bound by a periplasmic binding protein before being brought to a transporter in the inner membrane (Andrews et al. 2003; Cornelis 2010; Cornelis et al. 2011). Once in the cytoplasm, iron is liberated from the siderophore by a reductase or by enzymatic destruction of the siderophore molecule (Andrews et al. 2003). As we are going to see, this general scheme differs sometimes in the fluorescent pseudomonads in terms of siderophore-mediated iron uptake (Cornelis 2010).

3.2.1 Pyoverdine-Mediated Iron Uptake

All fluorescent pseudomonads produce the yellow-green fluorescent pyoverdine molecule, which is a strong iron chelator the purpose of which is to serve as the major siderophore for these bacteria (Cornelis 2010) (Fig. 3.1). These complex siderophores are synthesized via a non-ribosomal peptide synthesis pathway involving non-ribosomal peptide synthetases (NRPS) (Ravel and Cornelis 2003). Pyoverdines are composed of three parts, a chromophore, which confers the characteristic yellow-green color (and the fluorescence under UV exposure), a variable peptide chain of 6–12 amino acids, and a side chain, generally a di-carboxylic acid or a di-carboxylic amide (Meyer 2000; Mossialos et al. 2002; Ravel and Cornelis 2003). The peptide chain amino acids can be in D- or L-form, which confers resistance to proteases and some residues are modified, as for example the N^5-formyl-N^5-hydroxyornithine present in the pyoverdine from *P. aeruginosa* PAO1 (see Meyer, 2000 for a review). The precursor of pyoverdine, ferribactin, which is synthesized in the cytoplasm, is non-fluorescent and chromophore maturation takes place in the periplasm after transport of the pyoverdine precursor from the cytoplasm (Baysse et al. 2002; Yeterian et al. 2010). Recently, it was demonstrated that this precursor is myristoylated and that the acyl chains are removed by the acylase PvdQ (Hannauer et al. 2012b) (Fig. 3.1b shows the structure of the myristoylated ferribactin precursor without the peptide chain). Some pyoverdine biosynthesis enzymes, such as the L-ornithine N^5-oxygenase PvdA, are associated with the membrane and cluster at the old cell pole (Imperi et al. 2008; Guillon et al. 2012) while other biosynthetic enzymes, such as the PvdQ acylase do not show such clustering (Guillon et al. 2012). The NRPS responsible for the synthesis of pyoverdine in PAO1 are PvdL, PvdI, PvdJ, and

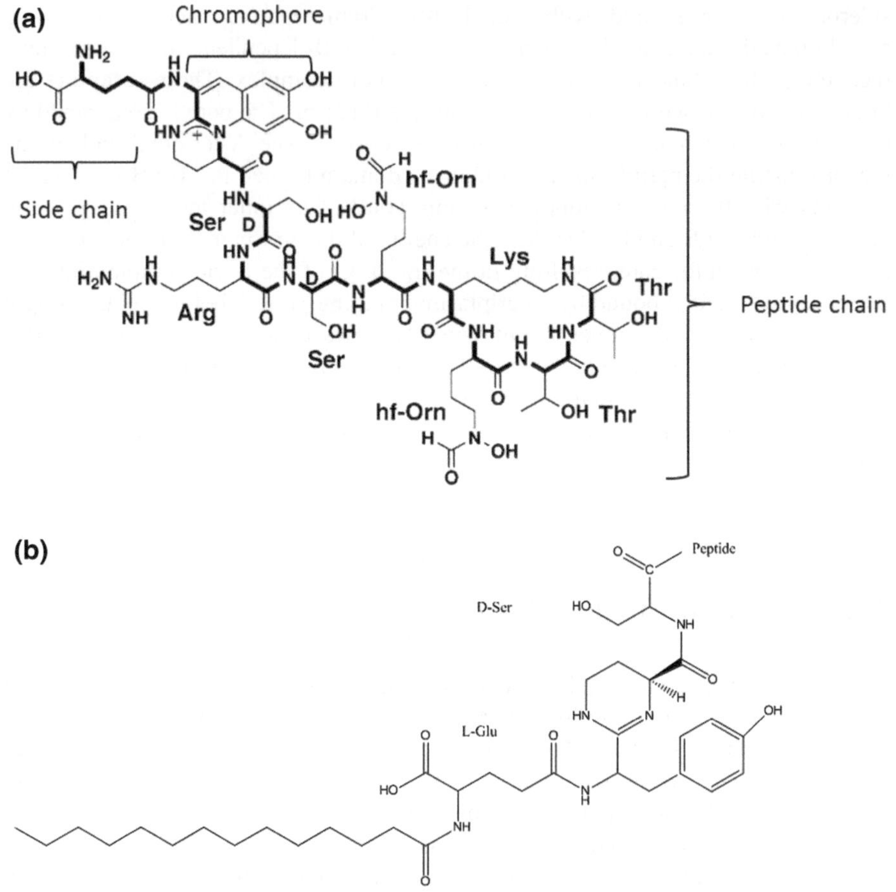

Fig. 3.1 **a** structure of the *P. aeruginosa* pyoverdine with the chromophore, the side chain, and the peptide chain (the peptide backbone is highlighted in thick line). The side chain can be either a succinamide, succinic acid, or α ketoglutaric acid (adapted from Ravel and Cornelis 2003). **b** structure of the myristoylated ferribactin chromophore precursor without the peptide chain (adapted from Hannauer et al. 2012b)

PvdD, which are large multi-modular enzymes where each module is responsible for the incorporation of one amino acid into the peptide chain. The activation (A)-domain of each module of the NRPS recognizes and activates a specific amino acid by reaction with ATP. This activated ester is then covalently linked as its thioester on the thiolation domain (**T**). The condensation domain (**C**) catalyzes the direct transfer to another acylamino acid intermediate on the adjacent downstream module to form a peptide bond. In some cases, epimerization of amino acid from the L- to the D-configurations is catalyzed by an extra domain (**E**). A terminal thioesterase releases the peptide from the enzyme by cyclization or hydrolysis. Only one NRPS, encoded by *pvdL* (PA2424), is highly conserved in all

Pseudomonas genomes analyzed since it is involved in the biosynthesis of the precursor of the chromophore, which is identical in almost all pyoverdines described so far (Mossialos et al. 2002). PvdL and the downstream NRPSs work in concert to synthesize the entire pyoverdine peptide backbone. As can be expected, given the large variety of pyoverdine peptide moiety structure (Meyer 2000), the downstream NRPSs do not show high degrees of similarity between fluorescent *Pseudomonas* species. Each pyoverdine is therefore species-specific, with the exception of *P. aeruginosa* where three different types of pyoverdine have been described (Meyer et al. 1997; Smith et al. 2005; Bodilis et al. 2009). This diversity in peptide chains of pyoverdines implies that each pyoverdine is recognized by a specifically interacting outer membrane receptor, suggesting a co-evolution between the modular NRPS enzymes and the receptor (de Chial et al. 2003; Smith et al. 2005; Tummler and Cornelis 2005; Bodilis et al. 2009).

The mechanism of pyoverdine export, ferripyoverdine uptake, and the release of iron in the periplasm is depicted in Fig. 3.2 (adapted from Cornelis, 2010). The acylated precursor of pyoverdine (Fig. 3.1b) is probably transported out of the cytoplasm by the ABC transporter PvdE (McMorran et al. 1996). Once in the

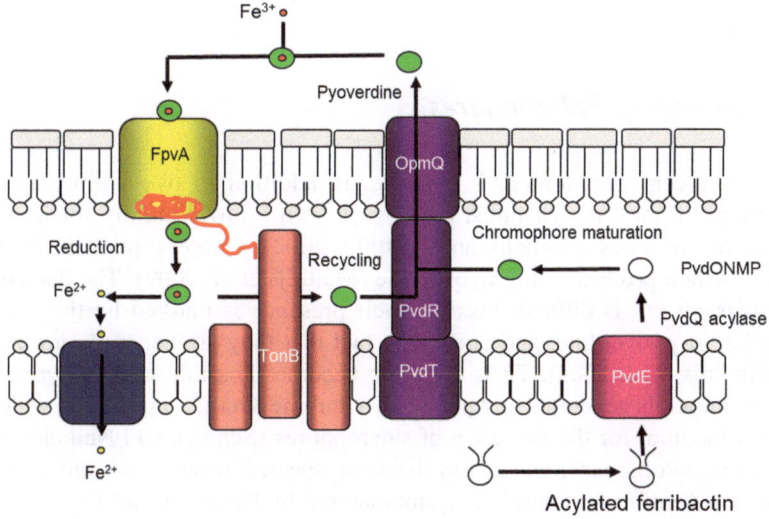

Fig. 3.2 Schematic representation of the pyoverdine-mediated iron uptake in fluorescent pseudomonads. The acylated ferribactin precursor is made in the cytoplasm and is transported to the periplasm via the PvdE ABC transporter where the PvdQ acylase removes the lipid tail while the PvdONMP proteins are probably involved in the maturation of the chromophore. The PvdTR-OpmQ transporter is involved in the transport of de novo synthesized pyoverdine and in the recycling of pyoverdine after reduction of Fe^{3+} to Fe^{2+}. Outside of the cell, pyoverdine will form a 1:1 complex with Fe^{3+} and the ferripyoverdine is recognized at the level of the outer membrane by the FpvA receptor. The receptor gets the energy from the inner membrane proton motive force via the TonB protein together with ExbB and ExbD to remove the cork (shown in red). In the periplasm the ferripyoverdine is immediately reduced in Fe^{2+} and apo-pyoverdine recycled via PvdTR-OpmQ

periplasm, the PvdQ acylase removes the acyl chains, releasing the non-fluorescent ferribactin precursor, which is later matured to form the pyoverdine with the fluorescent chromophore (Nadal Jimenez et al. 2010). Pyoverdine is exported outside of the cell by a mechanism involving, the PvdT-PvdR-OpmQ ABC export system. However, another efflux system may compensate for the absence of the pump as illustrated by Hannauer et al. (Hannauer et al. 2010). Once inside the periplasmic space, iron is removed from the pyoverdine siderophore by a still undescribed mechanism. Ferripyoverdine is unlikely to enter the cytoplasm. The recycling of pyoverdine via the PVDRT-OpmQ pump is fast according to kinetic studies, suggesting that pyoverdine does not enter the cytoplasm (Schalk et al. 2002; Greenwald et al. 2007; Imperi et al. 2009; Hannauer et al. 2012a). Recently, it has been shown that PvdRT-OpmQ is also involved in the re-export of unwanted metal-pyoverdine chelates, avoiding the accumulation of pyoverdines complexed with other metals than iron, which could be toxic for the cell (Hannauer et al. 2012a).

The uptake system described here for *P. aeruginosa* is probably very similar for other fluorescent pseudomonads, which share the same biosynthesis and uptake genes, although the genes are not always contiguous, with the exception of *P. syringae* (Ravel and Cornelis 2003).

3.2.2 Secondary Siderophores

Many fluorescent pseudomonads produce, in addition to pyoverdine, a second siderophore of various structures, but always with a lower affinity for iron compared to pyoverdines (Cornelis and Matthijs 2002). One exception is *P. putida* KT2440, which produces only pyoverdine (Matthijs et al. 2009). The detection of these siderophores is difficult because their presence is masked by the abundant production of pyoverdine, and due to the fact that they are generally not fluorescent (Matthijs et al. 2009). Their detection becomes possible when a pyoverdine-negative mutant is generated and plated on a chrome azurol S agar plate, which is a universal medium for the detection of siderophores (Schwyn and Neilands 1987). Different pseudomonads, even from the same species, produce secondary siderophores, which differ structurally, as summarized in Table 3.1 and Fig. 3.3.

Some of these siderophores with lower affinity for iron have interesting characteristics besides their capacity to bind iron (Cornelis and Matthijs 2002). This is the case for the *P. putida* (and also *P. stutzeri*, a non-fluorescent pseudomonads) siderophore PDTC (pyridine-2,6-bis monothiocarboxylic acid), which is able to degrade carbon tetrachloride, and to bind other metals than iron (Lewis et al. 2004; Leach et al. 2007). Pyochelin has redox activity (see below) and is able to degrade the toxic compound organotin via the production of hydroxyl radicals (Sun and Zhong 2006; Sun et al. 2006). Another example is the siderophore thioquinolobactin, which shows antagonism against the phytopathogenic Oomycete *Pythium*, while its spontaneous degradation product, quinolobactin, keeps its siderophore

Table 3.1 List of secondary siderophores produced by fluorescent pseudomonads

Siderophore	Species	Reference
Pyochelin	*P.aeruginosa* *Burkholderia cepacia*	(Cox et al. 1981; Farmer and Thomas 2004)
Enantio-pyochelin	*P. fluorescens*	(Youard et al. 2007)
Pseudomonine	*P. fluorescens* *P. entomophila*	(Mercado-Blanco et al. 2001; Matthijs et al. 2009)
Ornicorrugatin	*P. fluorescens*	(Matthijs et al. 2008)
(Thio)quinolobactin	*P. fluorescens*	(Mossialos et al. 2000; Matthijs et al. 2004; Matthijs et al. 2007)
Yersiniabactin	*P. syringae*	(Bultreys et al. 2006; Jones et al. 2007)
Pyridine-2,6-bis (thiocarboxylic acid) (PDTC)	*P. putida* *P. stutzeri*	(Lewis et al. 2004; Morales and Lewis 2006; Leach et al. 2007)
Achromobactin	*P. syringae*	(Berti and Thomas 2009; Owen and Ackerley 2011; Greenwald et al. 2012)

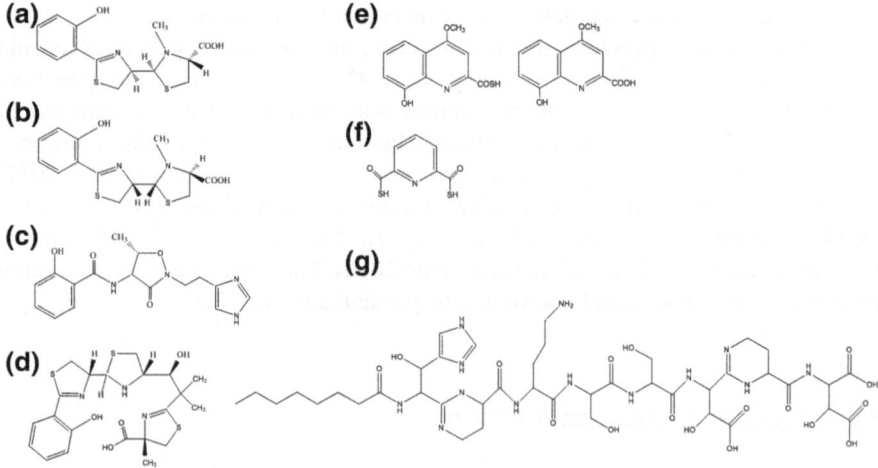

Fig. 3.3 Examples of structures of different Pseudomonas secondary siderophores **a** pyochelin from *P. aeruginosa*, **b** enantio-pyochelin from *P. fluorescens* Pf5 and CHA0, **c** pseudomonine from *P. fluorescens* and *P. entomophila*, **d** yersiniabactin from *P. syringae*, **e** thioquinolobactin and quinolobactin from *P. fluorescens*, **f** PDTC from *P. stutzeri* and *P. putida*, and **g** ornicorrugatin from *P. fluorescens*. The list of secondary siderophores is given in Table 3.1

activity but is inactive against *Pythium* (Mossialos et al. 2000; Matthijs et al. 2004; Matthijs et al. 2007). We will present here more in detail the interesting case of the *P. aeruginosa* pyochelin and its enantiomer, enantiopyochelin, produced by *P. fluorescens* Pf5 (Youard et al. 2011).

Pyochelin is a secondary siderophore produced by *P. aeruginosa*, which has a lower affinity for iron compared to pyoverdine (Cox et al. 1981) (Fig. 3.3). Pyochelin is produced by all *P. aeruginosa* isolates and, although it is not as important for the virulence of the bacterium as pyoverdine (Meyer et al. 1996), it can cause inflammation via the production of hydroxyl radicals, especially in combination with the extracellular phenazine pyocyanin (Coffman et al. 1990; Britigan et al. 1992; Britigan et al. 1997). Pyochelin is synthesized from salicylic acid and cysteine as precursors (Ankenbauer and Cox 1988; Serino et al. 1995; Serino et al. 1997). The ferripyochelin complex is recognized at the level of the outer membrane by the FptA TonB-dependent receptor (Ankenbauer 1992). Pyochelin makes a 2:1 complex with iron and the structure of the FptA receptor in complex with pyochelin has been solved (Cobessi et al. 2005; Mislin et al. 2006). Pyochelin being a lower affinity siderophore is able to chelate other metals besides iron, a property that has been used to probe the interaction of pyochelin-terbium with FptA (Yang et al. 2011).

Enantiopyochelin is a stereoisomer of pyochelin and is produced by *P. fluorescens* Pf5 and CHA0 (Fig. 3.3b) (Youard et al. 2007; Youard et al. 2011). Being enantiomers, each siderophore is recognized by a different outer membrane receptor, the already mentioned FptA for ferri-pyochelin and FetA for enantio-pyochelin (Youard et al. 2011) (Fig. 3.4). The same specificity is observed for the transport of the ferri-siderophores through the periplasm and the inner membrane (Reimmann 2012). In the case of *P. aeruginosa* pyochelin, there is no periplasmic-binding protein involved, but only one transporter in the inner membrane, FptX. FptX is able to transport both pyochelin and enantio-pyochelin in complex with iron; however, the transport of enantio-pyochelin involves both a periplasmic binding protein and an ABC transporter showing a strict specificity towards this stereoisomer form (Reimmann 2012) (Fig. 3.4). Another example of specificity is found in the AraC PchR regulator which is specific for pyochelin in the case of *P. aeruginosa* and for enantio-pyochelin in the case of *P. fluorescens* Pf5 (Youard and Reimmann 2010). The details of the regulation of siderophore biosynthesis and uptake will be presented in Sect. 3.5.

3.2.3 Uptake of Xenosiderophores

Fluorescent pseudomonads have mechanisms to take up a wide range of sidero-phores which they do not produce themselves (termed xenosiderophores) via dedicated TonB-dependent receptors (Cornelis and Matthijs 2002; Bodilis et al. 2009; Cornelis and Bodilis 2009; Cornelis 2010). Some receptors for xenosider-ophores have been characterized in *P. aeruginosa* (Table 3.2).

In other fluorescent pseudomonads, the number of TonB-dependent receptors can be quite high. For instance, *P. fluorescens* Pf5 has 45 genes for receptors that may allow the uptake of exogenous and heterologous pyoverdines (Hartney et al. 2011). The different TonB-dependent receptors can be classified into two categories, the simple TonB-dependent receptors (**TBDR**s, such as FpvB of *P. aeruginosa*, which is

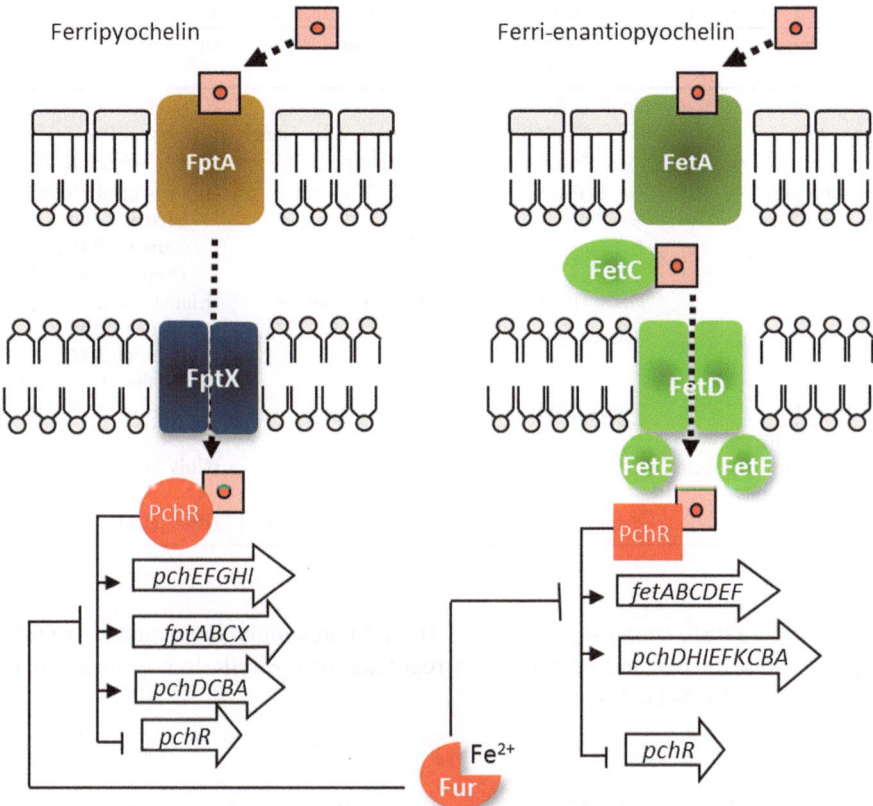

Fig. 3.4 Uptake and regulation of ferripyochelin (*left*) and ferrienantio-pyochelin (*right*). The ferripyochelin is taken up via the FptA receptor and transported to the cytoplasm via the FptX transporter while the ferrienantio-pyochelin is transported via the FetA receptor, the FetC periplasmic binding protein, and the FetDE ABC transporter. The PchR AraC regulators are also each specific for each type of enantiomers and positively regulate the biosynthesis genes, the receptor gene, and negatively regulate their own expression. The *pchR* genes are negatively regulated by Fur-Fe^{2+}

regulated solely by Fur), and the TonB-dependent transducers (**TBDT**) (Hartney et al. 2011). The TBDT can sense the presence of the cognate ferrisiderophore. Upon recognition, the TBDT *N*-terminal extension interacts with an anti-sigma factor in the inner membrane. The anti-sigma factor becomes prone to proteolysis and liberates an extracellular sigma factor (ECF s), which associates with the RNA polymerase to transcribe the receptor gene, causing an auto-induction reaction (the receptor is made only when the cognate ferrisiderophore is present) (Llamas et al. 2006; Cornelis et al. 2009; Mettrick and Lamont 2009; Draper et al. 2011; Hartney et al. 2011). In *P. fluorescens* Pf5, there are 27 TBDRs and 18 TBDT (Hartney et al. 2011). Recently, Elias et al. (Elias et al. 2011) described the FvbA receptor (PA4156), which allows the utilization of the xenosiderophore vibriobactin

Table 3.2 List of characterized xenosiderophore receptors in *P. aeruginosa*

Siderophore	Receptor and gene number (PAO1)	Regulation	Reference
Pyoverdine	FpvB (PA4198)	Fur	(Ghysels et al. 2004)
Enterobactin	PfeA (PA2688) PirA (PA0931)	Fur-two component system	(Dean and Poole 1993a; Ghysels et al. 2005)
Ferrichrome	FiuA (PA0470)	Fur-ECF σ-anti σ	(Llamas et al. 2006; Mettrick and Lamont 2009; Draper et al. 2011)
Ferrioxamine A	FoxA (PA2466)	Fur-ECF σ-anti σ	(Llamas et al. 2006; Mettrick and Lamont 2009; Draper et al. 2011)
Mycobactin, carboxymycobactin	FemA (PA1910)	Fur-ECF σ-anti σ	(Llamas et al. 2008)
Aerobactin, rhizobactin, schizokinen	ChtA (PA4675)	Fur	(Cuiv et al. 2006)
Vibriobactin	FvbA (PA4156)	Fur-PA4157 (FvbR)	(Elias et al. 2011)

produced by *Vibrio cholerae* (Elias et al. 2011). Interestingly, the regulation of *fvbA* depends both on Fur and on the FvbR regulator of the IclR family encoded by PA4157 (see also Sect. 3.5).

3.3 Heme Uptake in Fluorescent Pseudomonads

Like other bacteria, all fluorescent pseudomonads can take heme from hemoproteins released in the environment using different heme uptake systems, each involving a TonB-dependent receptor (Cornelis and Bodilis 2009). Heme must first be extracted from hemoproteins such as hemoglobin or hemopexin. Heme is not found in the unbound form because in its free form it is both toxic as well as highly hydrophobic causing it to associate with membranes where it promotes non-enzymatic redox reactions (Wyckoff et al. 2005). In *P. aeruginosa* there are two characterized heme uptake systems. The Has system uses the type I secretion of an extracellular heme binding protein (termed the hemophore), HasA, and the TonB-dependent receptor HasR. The Phu system uses only the TonB-dependent receptor, PhuR (Letoffe et al. 1998; Ochsner et al. 2000a). A third system, not yet characterized, involves the TonB-dependent receptor HxuC, which interestingly appears to be the most conserved TBDR among fluorescent pseudomonads (Cornelis and Bodilis 2009). In the periplasm, heme is bound by a heme-binding protein, and, once in the cytoplasm, it is broken down by a heme oxygenase, HemO, resulting in CO and biliverdin production (Fig. 3.5). Because of its toxicity, heme is not only bound to a periplasmic binding protein, but also to a cytoplasmic heme chaperone, PhuS, which forms a 1:1 complex

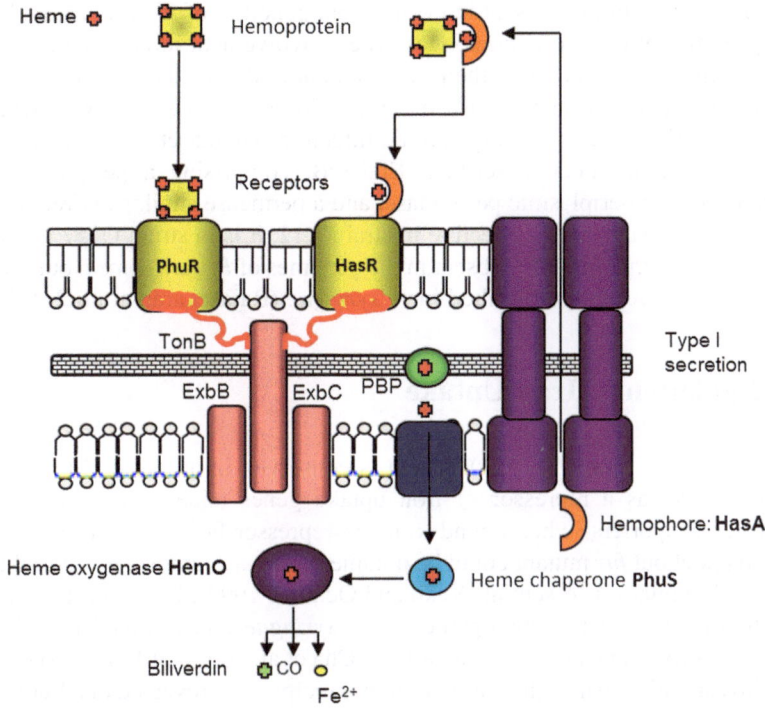

Fig. 3.5 Heme uptake systems in *P. aeruginosa*. The two described heme uptake systems presented are Phu and Has. Hemoproteins such as hemoglobin can either bind directly to the PhuR outer membrane receptor or a protein, HasA, which is first secreted by a type I secretion system to the outside where it can extract heme from hemoproteins. In the periplasm heme is bound to a periplasmic binding protein and is transported to the cytoplasm via an inner membrane transporter. Once in the cytoplasm, heme is bound by the heme chaperone PhuS, which directs it to the heme oxygenase, HemO. Heme is degraded by HemO into biliverdin, Co and Fe^{2+}

with the HemO heme oxygenase when in the holo form (O'Neill et al. 2012). A homolog of *P. aeruginosa* HasR is only found in the genome of the insect pathogen *P. entomophila*, suggesting that this heme uptake system is important for animal pathogens. Likewise, *P. entomophila* has a putative heme receptor with similarity to *P. aeruginosa* HxuC, which means that *P. entomophila*, like *P. aeruginosa*, has three heme uptake systems (Cornelis and Bodilis 2009).

3.4 Uptake of Fe^{2+}

Many bacteria are also able to take up Fe^{2+} via the Feo system (Cartron et al. 2006). Fe^{2+} is soluble and probably diffuses through the outer membrane and is further transported inside the cytoplasm by the FeOAB system (Cartron et al.

2006). The *feo* genes are present in the different genomes of *P. aeruginosa*, but not in the genomes of other pseudomonads. Their involvement in Fe^{2+} uptake has not been investigated yet, although their involvement in the uptake of iron could be of importance, especially in the case of cystic fibrosis. *P. aeruginosa* phenazines could reduce Fe^{3+} to Fe^{2+} during chronic infection (Hunter et al. 2012). In *E. coli* another system has been described, EfeUOB, comprising a periplasmic Fe^{2+} binding protein, a periplasmic peroxidase, and a permease similar to Fret1 of yeast and while this system is not effective in strain K12, it is in strain O157 (Cao et al. 2007). This system is however absent in the genomes of *P. aeruginosa*, but a similar system has been recently described in *P. syringae* (Rajasekaran et al. 2010).

3.5 Regulation of Iron Uptake

The ferric uptake regulator (Fur), which is conserved in different Gram-negative bacteria, works as a repressor of iron uptake genes (siderophore biosynthesis, receptors, transporters) when bound to its co-repressor Fe^{2+} (Escolar et al. 1999). Since no knockout *fur* mutant could be obtained in *P. aeruginosa*, it is possible that this central regulator is essential (Vasil and Ochsner 1999). However, a *fur* mutant could be obtained in the phytopathogen *P. syringae* and it was found that Fur regulates quorum sensing in this bacterium (Cha et al. 2008). Still, in *P. syringae*, a recent investigation using chromatin immunoprecipitation revealed that Fur binds to more than 300 regions in the genome and electrophoretic mobility shift assay (EMSA) confirmed the direct binding of Fur to 41 targets (Butcher et al. 2011). The Fur regulon in *P. aeruginosa* has also been investigated using a bioinformatics approach. This approach confirmed previously identified Fur-regulated genes and predicted new Fur targets (van Oeffelen et al. 2008; Cornelis et al. 2009). In fluorescent pseudomonads some genes are directly regulated by Fur, but the majority are indirectly regulated with Fur being at the top of the hierarchy of regulators (Cornelis et al. 2009).

3.5.1 ECF Sigma Factors

Fur indirect control involves extracytoplasmic sigma factors (ECF σ) or works via other regulators (Visca et al. 2007; Cornelis et al. 2009). The best investigated ECF σ is PvdS which not only controls the transcription of pyoverdine biosynthesis genes in *P. aeruginosa*, but also of virulence genes such as the gene encoding exotoxin A or the extracellular protease PrpL (Visca et al. 2007). Similar to the situation in *P. aeruginosa*, PvdS is involved in the transcription of numerous genes in *P. syringae*, including the genes involved in the biosynthesis of pyoverdine (Cornelis 2008; Swingle et al. 2008). By analyzing the PvdS binding motif of different PvdS-controlled genes in *P. syringae* and *P. aeruginosa*, Swingle et al. could predict the PvdS regulatory network. This study concluded that next to a conserved "core" regulon that included pyoverdine biosynthesis genes, PvdS has an "accessory" regulon which differs from species to species (Cornelis 2008; Swingle et al. 2008). In *P. aeruginosa*, there is a second ECF σ, FpvI, which is needed for the transcription of the *fpvA* gene encoding the ferripyoverdine receptor (Visca et al. 2007). In *P. aeruginosa* PAO1 genome there are 19 genes coding for ECF sigma factors (Potvin et al. 2008), 10 of which are known or predicted to be Fur-regulated (Llamas et al. 2008; van Oeffelen et al. 2008; Cornelis et al. 2009).

Generally, the genes encoding the ECF σ factors are adjacent to one gene encoding a transmembrane anti- σ factor and a gene encoding a TonB-dependent receptor for the uptake of a xenosiderophore, which they regulate according to the *E. coli* FecI/FecR paradigm described by Braun (Braun et al. 2006; Cornelis et al. 2009). This organization is important for the cell-surface signaling involving the ferrisiderophore receptor (Visca et al. 2007; Llamas et al. 2008) as illustrated in Fig. 3.6.

Llamas et al. (2008) also demonstrated that two ECF σ (PA0149 and PA2050) are necessary for the transcription of the PA2384 gene encoding a Fur-like regulator, which is important for the expression of iron uptake genes, including the pyoverdine, pyochelin, and heme uptake genes (Zheng et al. 2007). Inactivation of the PA2384 regulator also results in a higher expression of the genes for the biosynthesis of the quorum sensing signal molecule PQS (Pseudomonas quinolone signal) (Zheng et al. 2007).

3.5.2 Other Fur-Regulated Regulators

In *P. aeruginosa*, the *fptA* gene encoding the ferripyochelin receptor is under the control of the AraC regulator PchR (Michel et al. 2005). Fur is also predicted to repress the expression of PA3269, which encodes a putative AraC regulator and which is next to PA3268 coding for a TonB-dependent receptor (van Oeffelen et al. 2008; Cornelis et al. 2009). Two genes, *pirA* (PA0931) and *pfeA* (PA2688) each encoding a ferrienterobactin receptor (Ghysels et al. 2005), are regulated by two-component systems, their respective response regulators being PA0929 and PA2686 (*pfeR*) (Dean and Poole 1993a; Dean and Poole 1993b).

A list of other regulator genes predicted to be regulated by Fur-Fe^{2+} is given in Table 3.3.

3.5.3 Regulation by Small RNAs

Fur was found to regulate the transcription of two small RNA genes, PrrF1 and PrrF2 (Wilderman et al. 2004; Oglesby et al. 2008). Several genes which are induced by high iron conditions are also de-repressed in a Δ*prrF1,2* mutant grown in low iron conditions (Oglesby et al. 2008). Among them is the gene encoding the

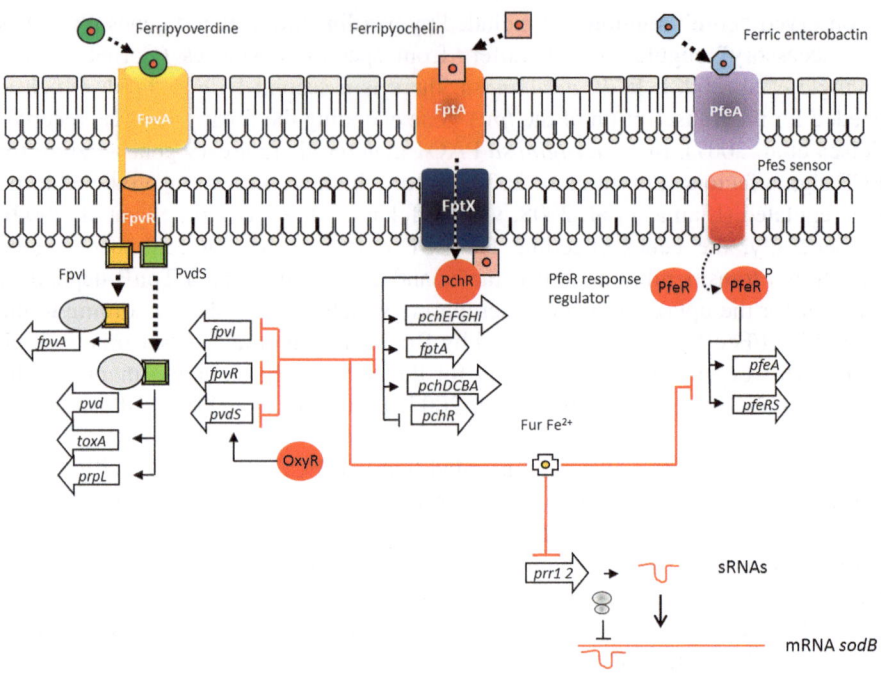

Fig. 3.6 Regulation of ferripyoverdine, ferripyochelin, and ferric enterobactin uptake in *P. aeruginosa*. Ferripyoverdine binds to the FpvA receptor, which triggers a conformational change and the proteolytic cleavage of the anti-sigma factor FpvR. This releases the ECF σ PvdS and FpvI, which can associate with the RNA polymerase to transcribe the pyoverdine biosynthesis genes (PvdS) or the *fpvA* gene (FpvI). PvdS is also needed for the transcription of the *toxA* gene encoding the exotoxin A and the *prpL* gene encoding an extracellular protease. Ferripyochelin binds to the FptA receptor and is transported inside the cell via the FptX transporter in the inner membrane. Pyochelin binds the the AraC PchR regulator which activates the transcription of the pyochelin biosynthesis and uptake genes while it acts as a repressor of its own expression. Ferric enterobactin binds to the PfeA receptor and is detected at the level of the inner membrane by the PfeS sensor protein which undergoes an auto-phosphorylation. The sensor phosphorylates in turn the response regulator PfeR which activates the expression of the *pfeA* and *pfeRS* genes. The same kind of two-component system is used to transport ferric enterobactin via the alternative PirA receptor (not shown). Fur-Fe^{2+} represses the transcription of *pvdS*, *fpvI*, *pchR* and *pfeA pfeRS*. Fur-Fe^{2+} also represses the expression of the *prr1 prr2* RNAs which work as anti-sense RNAs repressing the translation of some RNA. OxyR was also found to activate the transcription of *pvdS*

Fe-superoxide dismutase (*sodB*), Fe-aconitase A (*acnA*) and succinate dehydrogenase (*sdhCDAB*), as well as other TCA cycle enzymes encoding genes (Vasil 2007). Interestingly, among the genes that are regulated by these two small RNAs are the *antAB* and *catAB* genes for the degradation of anthranilate (Oglesby et al. 2008). Anthranilate is the precursor of the quinolone molecule, 2-heptyl-3-hydroxy-4-quinolone, also termed the Pseudomonas quinolone signal molecule (PQS), an important quorum sensing molecule (Diggle et al. 2006), clearly establishing a link between quorum sensing, virulence and iron regulation.

Table 3.3 List of *P. aeruginosa* known and predicted Fur-regulated regulator genes

Regulator gene	Type of regulator	Probable target	Reference
PA1315	TetR	PA1313 and PA1316 are major facilitator drug export genes (MFS)	(van Oeffelen et al. 2008; Cornelis et al. 2009)
PA1570	LysR	PA1569 (MFS)	(van Oeffelen et al. 2008; Cornelis et al. 2009)
PA1630	IclR	PA1631 (acyl-CoA dehydrogenase)	(van Oeffelen et al. 2008; Cornelis et al. 2009)
PA2056	LysR	PA2055 (MFS) PA2057 (TonB-dependent receptor)	(van Oeffelen et al. 2008; Cornelis et al. 2009)
PA2384	Fur-like	Iron homeostasis and quorum sensing	(Zheng et al. 2007)
PA3133	AcrR	PA3132 (hydrolase)	(van Oeffelen et al. 2008; Cornelis et al. 2009)
PA4157(*fvbR*)	IclR	PA4156 (FvbA)	(Elias et al. 2011)
PA4296 (*pprB*)	Response	regulator	PA4293 encodes the sensor PprA
(van Oeffelen et al. 2008; Cornelis et al. 2009)			
PA4315 (mvaT)	H-NS	Global regulator, type IV pili, quorum sensing	(Vallet et al. 2004)
PA5437	LysR	PA5436 (biotin carboxylase)	(van Oeffelen et al. 2008; Cornelis et al. 2009)

3.5.4 Other Regulators Influencing Iron Homeostasis

Quorum sensing (QS) is a way by which bacteria respond to the presence of signal molecules, which they produce upon achieving a certain high density, allowing them to coordinately adapt their behavior. Detection of the quorum sensing molecules triggers the expression of a large number of genes, including colonization and virulence genes (Bassler 1999; Camilli and Bassler 2006; Ng and Bassler 2009). In addition to the production of virulence factors, QS in *P. aeruginosa* also regulates several other cellular processes including biofilm formation, adhesion, genes involved in the oxidative stress response, osmotic stress, cold shock, and exposure to antibiotics (Hentzer et al. 2003; Schuster et al. 2003; Wagner et al. 2003; Gilbert et al. 2009). Three interconnected QS systems have been identified as a major gene regulatory circuit in *P. aeruginosa* (Venturi 2006; Williams and Camara 2009). The LasR/LasI and RhlR/RhlI systems, produce *N*-3-(oxododecanoyl) homoserine lactone (3-oxo-C12-HSL) and the *N*-butanoyl homoserine lactone (C4-HSL), respectively, that orchestrate the expression of hundreds of genes (Venturi 2006; Williams and Camara 2009). *P. aeruginosa* produces diverse 4-quinolones QS signal molecules, among which the 2-heptyl-4-quinolone (HHQ) and 2-heptyl-3-hydroxy-4-quinolone (PQS), also known as Pseudomonas quinolone signals (PQS)

(Diggle et al. 2006; Heeb et al. 2011). PQS is also able to bind iron (Bredenbruch et al. 2006; Diggle et al. 2007), influencing the production of siderophores, although it is not itself a siderophore since PQS-Fe^{3+} is unable to sustain the growth of a *P. aeruginosa* mutant unable to produce the two siderophores, pyoverdine and pyochelin (Diggle et al. 2007). The *P. aeruginosa* QS LuxR regulator VqsR was involved in the regulation of virulence, but also in the regulation of siderophore production since pyoverdine and pyochelin genes expression was decreased in a *vqsR* mutant (Cornelis and Aendekerk 2004; Juhas et al. 2004). The PA2384 gene encodes a regulator with weak similarity to Fur. A PA2384 knockout mutant has decreased expression of iron uptake genes while the transcription of the PQS genes is enhanced (Zheng et al. 2007). The LysR regulator OxyR in *P. aeruginosa* is activated by exposure to H_2O_2 by the formation of a disulfide bridge, allowing it to activate the transcription of oxidative stress defense genes (encoding catalases, alkyl-hydro-peroxidases) (Ochsner et al. 2000b; Heo et al. 2010). We previously discovered that an *oxyR* mutant in *P. aeruginosa* is unable to utilize pyoverdine as a siderophore although it produces the siderophore and is able to take it up. One possibility is that the reduction of ferripyoverdine in the periplasm is not taking place (Vinckx et al. 2008). More recently, using chromatin immunoprecipitation, we found that the *pvdS* gene is a target of OxyR (Wei et al. 2012). Binding of purified oxidized OxyR to a DNA sequence upstream of *pvdS* was confirmed by electrophoretic mobility shift assay (EMSA) and expression of *pvdS* was found to be increased after exposure to H_2O_2 in the wild-type *P. aeruginosa*, but not in the *oxyR* mutant, confirming the role of OxyR in *P. aeruginosa* as an activator of *pvdS* expression (Wei et al. 2012).

3.6 Conclusions

Iron uptake and homeostasis form a critical issue for ubiquitous bacteria such as the fluorescent pseudomonads, which are able to thrive in very diverse environments. These bacteria face competition with other microorganisms and are able to colonize hosts as diverse as nematodes, insects, plants, or mammals. To illustrate this diversity we can point out the multiple pyoverdine siderophores, each characteristic of one species or even strains. The uptake and recycling of pyoverdines also shows differences with what is known for other siderophore uptake systems of Gram-negative bacteria. Also striking, is the diversity of secondary siderophores as well as their side activities, which can open biotechnological perspectives (such as the degradation of toxic compounds or their antagonistic activities against bacteria and fungi). The wide range and number of receptors for xenosiderophores is also a characteristic of fluorescent *Pseudomonas* while the uptake of heme seems to be more redundant and sophisticated in insect and human pathogens. It is therefore not surprising that the control of iron homeostasis is quite complicated, often involving several layers of regulators ranging from direct control by Fur to the involvement of surface signaling systems via ECF sigmas and anti-sigmas or other types of regulators, including post-transcriptional regulation by small RNAs.

More recently, it became clear that there is a cross-talk between quorum sensing, oxidative stress response, and iron homeostasis in fluorescent pseudomonads. Many questions remain unanswered such as the mechanism of reduction of ferrisiderophores, the involvement of many secondary regulators, and the extension of the study of iron uptake and homeostasis in fluorescent pseudomonads other than *P. aeruginosa*.

References

Abd H, Wretlind B, Saeed A, Idsund E, Hultenby K, Sandstrom G (2008) *Pseudomonas aeruginosa* utilises its type III secretion system to kill the free-living amoeba *Acanthamoeba castellanii*. J Eukaryot Microbiol 55:235–243

Andrews SC, Robinson AK, Rodriguez-Quinones F (2003) Bacterial iron homeostasis. FEMS Microbiol Rev 27:215–237

Ankenbauer RG (1992) Cloning of the outer membrane high-affinity Fe(III)-pyochelin receptor of *Pseudomonas aeruginosa*. J Bacteriol 174:4401–4409

Ankenbauer RG, Cox CD (1988) Isolation and characterization of *Pseudomonas aeruginosa* mutants requiring salicylic acid for pyochelin biosynthesis. J Bacteriol 170:5364–5367

Apidianakis Y, Rahme LG (2009) *Drosophila melanogaster* as a model host for studying *Pseudomonas aeruginosa* infection. Nat Protoc 4:1285–1294

Bassler BL (1999) How bacteria talk to each other: regulation of gene expression by quorum sensing. Curr Opin Microbiol 2:582–587

Baysse C, Budzikiewicz H, Uria Fernandez D, Cornelis P (2002) Impaired maturation of the siderophore pyoverdine chromophore in *Pseudomonas fluorescens* ATCC 17400 deficient for the cytochrome *c* biogenesis protein CcmC. FEBS Lett 523:23–28

Berti AD, Thomas MG (2009) Analysis of achromobactin biosynthesis by *Pseudomonas syringae* pv. syringae B728a. J Bacteriol 191:4594–4604

Bodilis J et al (2009) Distribution and evolution of ferripyoverdine receptors in *Pseudomonas aeruginosa*. Environ Microbiol 11:2123–2135

Braun V, Killmann H (1999) Bacterial solutions to the iron-supply problem. Trends Biochem Sci 24:104–109

Braun V, Mahren S, Sauter A (2006) Gene regulation by transmembrane signaling. Biometals 19:103–113

Bredenbruch F, Geffers R, Nimtz M, Buer J, Haussler S (2006) The *Pseudomonas aeruginosa* quinolone signal (PQS) has an iron-chelating activity. Environ Microbiol 8:1318–1329

Britigan BE, Rasmussen GT, Cox CD (1997) Augmentation of oxidant injury to human pulmonary epithelial cells by the *Pseudomonas aeruginosa* siderophore pyochelin. Infect Immun 65:1071–1076

Britigan BE, Roeder TL, Rasmussen GT, Shasby DM, McCormick ML, Cox CD (1992) Interaction of the *Pseudomonas aeruginosa* secretory products pyocyanin and pyochelin generates hydroxyl radical and causes synergistic damage to endothelial cells. Implications for Pseudomonas-associated tissue injury. J Clin Invest 90:2187–2196

Buell CR et al (2003) The complete genome sequence of the *Arabidopsis* and tomato pathogen *Pseudomonas syringae* pv. *tomato* DC3000. Proc Natl Acad Sci U S A 100:10181–10186

Bultreys A, Gheysen I, de Hoffmann E (2006) Yersiniabactin production by *Pseudomonas syringae* and *Escherichia coli*, and description of a second yersiniabactin locus evolutionary group. Appl Environ Microbiol 72:3814–3825

Butcher BG et al (2011) Characterization of the Fur regulon in *Pseudomonas syringae* pv. tomato DC3000. J Bacteriol 193:4598–4611

Camilli A, Bassler BL (2006) Bacterial small-molecule signaling pathways. Science 311: 1113–1116

Cao J, Woodhall MR, Alvarez J, Cartron ML, Andrews SC (2007) EfeUOB (YcdNOB) is a tripartite, acid-induced and CpxAR-regulated, low-pH Fe^{2+} transporter that is cryptic in *Escherichia coli* K-12 but functional in *E. coli* O157:H7. Mol Microbiol 65:857–875

Cartron ML, Maddocks S, Gillingham P, Craven CJ, Andrews SC (2006) Feo transport of ferrous iron into bacteria. Biometals 19:143–157

Cha JY, Lee JS, Oh JI, Choi JW, Baik HS (2008) Functional analysis of the role of Fur in the virulence of *Pseudomonas syringae* pv. *tabaci* 11528: Fur controls expression of genes involved in quorum-sensing. Biochem Biophys Res Commun 366:281–287

Cobessi D, Celia H, Pattus F (2005) Crystal structure at high resolution of ferric-pyochelin and its membrane receptor FptA from *Pseudomonas aeruginosa*. J Mol Biol 352:893–904

Coffman TJ, Cox CD, Edeker BL, Britigan BE (1990) Possible role of bacterial siderophores in inflammation. Iron bound to the *Pseudomonas* siderophore pyochelin can function as a hydroxyl radical catalyst. J Clin Invest 86:1030–1037

Cornelis P (2008) The 'core' and 'accessory' regulons of Pseudomonas-specific extracytoplasmic sigma factors. Mol Microbiol 68:810–812

Cornelis P (2010) Iron uptake and metabolism in pseudomonads. Appl Microbiol Biotechnol 86:1637–1645

Cornelis P, Aendekerk S (2004) A new regulator linking quorum sensing and iron uptake in *Pseudomonas aeruginosa*. Microbiology 150:752–756

Cornelis P, Bodilis J (2009) A survey of TonB-dependent receptors in fluorescent pseudomonads. Environ Microbiol Reports 1:256–262

Cornelis P, Matthijs S (2002) Diversity of siderophore-mediated iron uptake systems in fluorescent pseudomonads: not only pyoverdines. Environ Microbiol 4:787–798

Cornelis P, Matthijs S, Van Oeffelen L (2009) Iron uptake regulation in *Pseudomonas aeruginosa*. Biometals 22:15–22

Cornelis P, Wei Q, Andrews SC, Vinckx T (2011) Iron homeostasis and management of oxidative stress response in bacteria. Metallomics 3:540–549

Cox CD, Rinehart KL Jr, Moore ML, Cook JC Jr (1981) Pyochelin: novel structure of an iron-chelating growth promoter for *Pseudomonas aeruginosa*. Proc Natl Acad Sci U S A 78: 4256–4260

Cuiv PO, Clarke P, O'Connell M (2006) Identification and characterization of an iron-regulated gene, *chtA*, required for the utilization of the xenosiderophores aerobactin, rhizobactin 1021 and schizokinen by *Pseudomonas aeruginosa*. Microbiology 152:945–954

de Chial M et al (2003) Identification of type II and type III pyoverdine receptors from *Pseudomonas aeruginosa*. Microbiology 149:821–831

Dean CR, Poole K (1993a) Cloning and characterization of the ferric enterobactin receptor gene (*pfeA*) of *Pseudomonas aeruginosa*. J Bacteriol 175:317–324

Dean CR, Poole K (1993b) Expression of the ferric enterobactin receptor (PfeA) of *Pseudomonas aeruginosa*: involvement of a two-component regulatory system. Mol Microbiol 8:1095–1103

Diggle SP, Cornelis P, Williams P, Camara M (2006) 4-quinolone signalling in *Pseudomonas aeruginosa*: old molecules, new perspectives. Int J Med Microbiol 296:83–91

Diggle SP et al (2007) The *Pseudomonas aeruginosa* 4-quinolone signal molecules HHQ and PQS play multifunctional roles in quorum sensing and iron entrapment. Chem Biol 14:87–96

Draper RC, Martin LW, Beare PA, Lamont IL (2011) Differential proteolysis of sigma regulators controls cell-surface signalling in *Pseudomonas aeruginosa*. Mol Microbiol 82:1444–1453

Elias S, Degtyar E, Banin E (2011) FvbA is required for vibriobactin utilization in *Pseudomonas aeruginosa*. Microbiology 157:2172–2180

Escolar L, Perez-Martin J, de Lorenzo V (1999) Opening the iron box: transcriptional metalloregulation by the Fur protein. J Bacteriol 181:6223–6229

Farmer KL, Thomas MS (2004) Isolation and characterization of *Burkholderia cenocepacia* mutants deficient in pyochelin production: pyochelin biosynthesis is sensitive to sulfur availability. J Bacteriol 186:270–277

Ghysels B et al (2004) FpvB, an alternative type I ferripyoverdine receptor of *Pseudomonas aeruginosa*. Microbiology 150:1671–1680

Ghysels B et al (2005) The *Pseudomonas aeruginosa pirA* gene encodes a second receptor for ferrienterobactin and synthetic catecholate analogues. FEMS Microbiol Lett 246:167–174

Gilbert KB, Kim TH, Gupta R, Greenberg EP, Schuster M (2009) Global position analysis of the *Pseudomonas aeruginosa* quorum-sensing transcription factor LasR. Mol Microbiol 73: 1072–1085

Goldberg JB (2000) *Pseudomonas*: global bacteria. Trends Microbiol 8:55–57

Greenwald J et al (2007) Real time fluorescent resonance energy transfer visualization of ferric pyoverdine uptake in *Pseudomonas aeruginosa*. A role for ferrous iron. J Biol Chem 282:2987–2995

Greenwald JW, Greenwald CJ, Philmus BJ, Begley TP, Gross DC (2012) RNA-seq analysis reveals that an ECF sigma factor, AcsS, regulates achromobactin biosynthesis in *Pseudomonas syringae* pv. *syringae* B728a. PLoS ONE 7:e34804

Guillon L, El Mecherki M, Altenburger S, Graumann PL, Schalk IJ (2012) High cellular organization of pyoverdine biosynthesis in *Pseudomonas aeruginosa*: clustering of PvdA at the old cell pole. Environ Microbiol 14:1982–1994

Hannauer M, Braud A, Hoegy F, Ronot P, Boos A, Schalk IJ (2012a) The PvdRT-OpmQ efflux pump controls the metal selectivity of the iron uptake pathway mediated by the siderophore pyoverdine in *Pseudomonas aeruginosa*. Environ Microbiol 14:1696–1708

Hannauer M et al (2012b) Biosynthesis of the pyoverdine siderophore of *Pseudomonas aeruginosa* involves precursors with a myristic or a myristoleic acid chain. FEBS Lett 586:96–101

Hannauer M, Yeterian E, Martin LW, Lamont IL, Schalk IJ (2010) An efflux pump is involved in secretion of newly synthesized siderophore by *Pseudomonas aeruginosa*. FEBS Lett 584:4751–4755

Hartney SL, Mazurier S, Kidarsa TA, Quecine MC, Lemanceau P, Loper JE (2011) TonB-dependent outer-membrane proteins and siderophore utilization in *Pseudomonas fluorescens* Pf-5. Biometals 24:193–213

Heeb S, Fletcher MP, Chhabra SR, Diggle SP, Williams P, Camara M (2011) Quinolones: from antibiotics to autoinducers. FEMS Microbiol Rev 35:247–274

Hentzer M et al (2003) Attenuation of *Pseudomonas aeruginosa* virulence by quorum sensing inhibitors. EMBO J 22:3803–3815

Heo YJ et al (2010) The major catalase gene (*katA*) of *Pseudomonas aeruginosa* PA14 is under both positive and negative control of the global transactivator OxyR in response to hydrogen peroxide. J Bacteriol 192:381–390

Hunter RC, Klepac-Ceraj V, Lorenzi MM, Grotzinger H, Martin TR, Newman DK (2012) Phenazine content in the cystic fibrosis respiratory tract negatively correlates with lung function and microbial complexity. Am J Respir Cell Mol Biol 10(3):216–22

Imperi F et al (2008) Membrane-association determinants of the omega-amino acid monooxygenase PvdA, a pyoverdine biosynthetic enzyme from *Pseudomonas aeruginosa*. Microbiology 154:2804–2813

Imperi F, Tiburzi F, Visca P (2009) Molecular basis of pyoverdine siderophore recycling in *Pseudomonas aeruginosa*. Proc Natl Acad Sci U S A 106:20440–20445

Joardar V et al (2005) Whole-genome sequence analysis of *Pseudomonas syringae* pv. *phaseolicola* 1448A reveals divergence among pathovars in genes involved in virulence and transposition. J Bacteriol 187:6488–6498

Jones AM, Lindow SE, Wildermuth MC (2007) Salicylic acid, yersiniabactin, and pyoverdin production by the model phytopathogen *Pseudomonas syringae* pv. *tomato* DC3000: synthesis, regulation, and impact on tomato and Arabidopsis host plants. J Bacteriol 189:6773–6786

Juhas M et al (2004) Global regulation of quorum sensing and virulence by VqsR in *Pseudomonas aeruginosa*. Microbiology 150:831–841

Leach LH, Morris JC, Lewis TA (2007) The role of the siderophore pyridine-2,6-bis (thiocarboxylic acid) (PDTC) in zinc utilization by *Pseudomonas putida* DSM 3601. Biometals 20:717–726

Letoffe S, Redeker V, Wandersman C (1998) Isolation and characterization of an extracellular haem-binding protein from *Pseudomonas aeruginosa* that shares function and sequence similarities with the *Serratia marcescens* HasA haemophore. Mol Microbiol 28:1223–1234

Lewis TA et al (2004) Physiological and molecular genetic evaluation of the dechlorination agent, pyridine-2,6-bis(monothiocarboxylic acid) (PDTC) as a secondary siderophore of *Pseudomonas*. Environ Microbiol 6:159–169

Llamas MA, Mooij MJ, Sparrius M, Vandenbroucke-Grauls CM, Ratledge C, Bitter W (2008) Characterization of five novel *Pseudomonas aeruginosa* cell-surface signalling systems. Mol Microbiol 67:458–472

Llamas MA, Sparrius M, Kloet R, Jimenez CR, Vandenbroucke-Grauls C, Bitter W (2006) The heterologous siderophores ferrioxamine B and ferrichrome activate signaling pathways in *Pseudomonas aeruginosa*. J Bacteriol 188:1882–1891

Mahajan-Miklos S, Tan MW, Rahme LG, Ausubel FM (1999) Molecular mechanisms of bacterial virulence elucidated using a *Pseudomonas aeruginosa-Caenorhabditis elegans* pathogenesis model. Cell 96:47–56

Matthijs S et al (2004) The *Pseudomonas* siderophore quinolobactin is synthesized from xanthurenic acid, an intermediate of the kynurenine pathway. Mol Microbiol 52:371–384

Matthijs S, Budzikiewicz H, Schafer M, Wathelet B, Cornelis P (2008) Ornicorrugatin, a new siderophore from *Pseudomonas fluorescens* AF76. Z Naturforsch C 63:8–12

Matthijs S et al (2009) Siderophore-mediated iron acquisition in the entomopathogenic bacterium *Pseudomonas entomophila* L48 and its close relative *Pseudomonas putida* KT2440. Biometals 22:951–964

Matthijs S, Tehrani KA, Laus G, Jackson RW, Cooper RM, Cornelis P (2007) Thioquinolobactin, a *Pseudomonas* siderophore with antifungal and anti-*Pythium* activity. Environ Microbiol 9:425–434

McMorran BJ, Merriman ME, Rombel IT, Lamont IL (1996) Characterisation of the *pvdE* gene which is required for pyoverdine synthesis in *Pseudomonas aeruginosa*. Gene 176:55–59

Mercado-Blanco J, van der Drift KM, Olsson PE, Thomas-Oates JE, van Loon LC, Bakker PA (2001) Analysis of the *pmsCEAB* gene cluster involved in biosynthesis of salicylic acid and the siderophore pseudomonine in the biocontrol strain *Pseudomonas fluorescens* WCS374. J Bacteriol 183:1909–1920

Mettrick KA, Lamont IL (2009) Different roles for anti-sigma factors in siderophore signalling pathways of *Pseudomonas aeruginosa*. Mol Microbiol 74:1257–1271

Meyer JM (2000) Pyoverdines: pigments, siderophores and potential taxonomic markers of fluorescent *Pseudomonas* species. Arch Microbiol 174:135–142

Meyer JM, Neely A, Stintzi A, Georges C, Holder IA (1996) Pyoverdin is essential for virulence of *Pseudomonas aeruginosa*. Infect Immun 64:518–523

Meyer JM et al (1997) Use of siderophores to type pseudomonads: the three *Pseudomonas aeruginosa* pyoverdine systems. Microbiol 143:35–43

Michel L, Gonzalez N, Jagdeep S, Nguyen-Ngoc T, Reimmann C (2005) PchR-box recognition by the AraC-type regulator PchR of *Pseudomonas aeruginosa* requires the siderophore pyochelin as an effector. Mol Microbiol 58:495–509

Mislin GL, Hoegy F, Cobessi D, Poole K, Rognan D, Schalk IJ (2006) Binding properties of pyochelin and structurally related molecules to FptA of *Pseudomonas aeruginosa*. J Mol Biol 357:1437–1448

Morales SE, Lewis TA (2006) Transcriptional regulation of the pdt gene cluster of *Pseudomonas stutzeri* KC involves an AraC/XylS family transcriptional activator (PdtC) and the cognate siderophore pyridine-2,6-bis(thiocarboxylic acid). Appl Environ Microbiol 72:6994–7002

Mossialos D et al (2000) Quinolobactin, a new siderophore of *Pseudomonas fluorescens* ATCC 17400, the production of which is repressed by the cognate pyoverdine. Appl Environ Microbiol 66:487–492

Mossialos D et al (2002) Identification of new, conserved, non-ribosomal peptide synthetases from fluorescent pseudomonads involved in the biosynthesis of the siderophore pyoverdine. Mol Microbiol 45:1673–1685

Nadal Jimenez P et al (2010) Role of PvdQ in *Pseudomonas aeruginosa* virulence under iron-limiting conditions. Microbiol 156:49–59

Nelson KE et al (2002) Complete genome sequence and comparative analysis of the metabolically versatile *Pseudomonas putida* KT2440. Environ Microbiol 4:799–808

Ng WL, Bassler BL (2009) Bacterial quorum-sensing network architectures. Annu Rev Genet 43:197–222

O'Neill MJ, Bhakta MN, Fleming KG, Wilks A (2012) Induced fit on heme binding to the *Pseudomonas aeruginosa* cytoplasmic protein (PhuS) drives interaction with heme oxygenase (HemO). Proc Natl Acad Sci U S A 109:5639–5644

Ochsner UA, Johnson Z, Vasil ML (2000a) Genetics and regulation of two distinct haem-uptake systems, *phu* and *has*, in *Pseudomonas aeruginosa*. Microbiol 146:185–198

Ochsner UA, Vasil ML, Alsabbagh E, Parvatiyar K, Hassett DJ (2000b) Role of the *Pseudomonas aeruginosa oxyR-recG* operon in oxidative stress defense and DNA repair: OxyR-dependent regulation of *katB-ankB*, *ahpB*, and *ahpC-ahpF*. J Bacteriol 182:4533–4544

Oglesby AG et al (2008) The influence of iron on *Pseudomonas aeruginosa* physiology: a regulatory link between iron and quorum sensing. J Biol Chem 283:15558–15567

Owen JG, Ackerley DF (2011) Characterization of pyoverdine and achromobactin in *Pseudomonas syringae* pv. *phaseolicola* 1448a. BMC Microbiol 11:218

Paulsen IT et al (2005) Complete genome sequence of the plant commensal *Pseudomonas fluorescens* Pf-5. Nat Biotechnol 23:873–878

Potvin E, Sanschagrin F, Levesque RC (2008) Sigma factors in *Pseudomonas aeruginosa*. FEMS Microbiol Rev 32:38–55

Pukatzki S, Kessin RH, Mekalanos JJ (2002) The human pathogen *Pseudomonas aeruginosa* utilizes conserved virulence pathways to infect the social amoeba *Dictyostelium discoideum*. Proc Natl Acad Sci U S A 99:3159–3164

Rajasekaran MB et al (2010) Isolation and characterisation of EfeM, a periplasmic component of the putative EfeUOBM iron transporter of *Pseudomonas syringae* pv. *syringae*. Biochem Biophys Res Commun 398:366–371

Ravel J, Cornelis P (2003) Genomics of pyoverdine-mediated iron uptake in pseudomonads. Trends Microbiol 11:195–200

Reimmann C (2012) Inner-membrane transporters for the siderophores pyochelin in *Pseudomonas aeruginosa* and enantio-pyochelin in *Pseudomonas fluorescens* display different enantioselectivities. Microbiology 158:1317–1324

Schalk IJ, Abdallah MA, Pattus F (2002) Recycling of pyoverdin on the FpvA receptor after ferric pyoverdin uptake and dissociation in *Pseudomonas aeruginosa*. Biochem 41:1663–1671

Schuster M, Lostroh CP, Ogi T, Greenberg EP (2003) Identification, timing, and signal specificity of *Pseudomonas aeruginosa* quorum-controlled genes: a transcriptome analysis. J Bacteriol 185:2066–2079

Schwyn B, Neilands JB (1987) Universal chemical assay for the detection and determination of siderophores. Anal Biochem 160:47–56

Serino L, Reimmann C, Baur H, Beyeler M, Visca P, Haas D (1995) Structural genes for salicylate biosynthesis from chorismate in *Pseudomonas aeruginosa*. Mol Gen Genet 249:217–228

Serino L, Reimmann C, Visca P, Beyeler M, Chiesa VD, Haas D (1997) Biosynthesis of pyochelin and dihydroaeruginoic acid requires the iron-regulated *pchDCBA* operon in *Pseudomonas aeruginosa*. J Bacteriol 179:248–257

Smith EE, Sims EH, Spencer DH, Kaul R, Olson MV (2005) Evidence for diversifying selection at the pyoverdine locus of *Pseudomonas aeruginosa*. J Bacteriol 187:2138–2147

Stover CK et al (2000) Complete genome sequence of *Pseudomonas aeruginosa* PAO1, an opportunistic pathogen. Nature 406:959–964

Sun GX, Zhong JJ (2006) Mechanism of augmentation of organotin decomposition by ferripyochelin: formation of hydroxyl radical and organotin-pyochelin-iron ternary complex. Appl Environ Microbiol 72:7264–7269

Sun GX, Zhou WQ, Zhong JJ (2006) Organotin decomposition by pyochelin, secreted by *Pseudomonas aeruginosa* even in an iron-sufficient environment. Appl Environ Microbiol 72:6411–6413

Swingle B, Thete D, Moll M, Myers CR, Schneider DJ, Cartinhour S (2008) Characterization of the PvdS-regulated promoter motif in *Pseudomonas syringae* pv. *tomato* DC3000 reveals regulon members and insights regarding PvdS function in other pseudomonads. Mol Microbiol 68:871–889

Tummler B, Cornelis P (2005) Pyoverdine receptor: a case of positive Darwinian selection in *Pseudomonas aeruginosa*. J Bacteriol 187:3289–3292

Vallet I et al (2004) Biofilm formation in *Pseudomonas aeruginosa*: fimbrial *cup* gene clusters are controlled by the transcriptional regulator MvaT. J Bacteriol 186:2880–2890

van Oeffelen L, Cornelis P, Van Delm W, De Ridder F, De Moor B, Moreau Y (2008) Detecting cis-regulatory binding sites for cooperatively binding proteins. Nucleic Acids Res 36:e46

Vasil ML (2007) How we learnt about iron acquisition in *Pseudomonas aeruginosa*: a series of very fortunate events. Biometals 20:587–601

Vasil ML, Ochsner UA (1999) The response of *Pseudomonas aeruginosa* to iron: genetics, biochemistry and virulence. Mol Microbiol 34:399–413

Venturi V (2006) Regulation of quorum sensing in *Pseudomonas*. FEMS Microbiol Rev 30: 274–291

Vinckx T, Matthijs S, Cornelis P (2008) Loss of the oxidative stress regulator OxyR in *Pseudomonas aeruginosa* PAO1 impairs growth under iron-limited conditions. FEMS Microbiol Lett 288:258–265

Visca P, Imperi F, Lamont IL (2007) Pyoverdine siderophores: from biogenesis to biosignificance. Trends Microbiol 15:22–30

Vodovar N et al (2006) Complete genome sequence of the entomopathogenic and metabolically versatile soil bacterium *Pseudomonas entomophila*. Nat Biotechnol 24:673–679

Wagner VE, Bushnell D, Passador L, Brooks AI, Iglewski BH (2003) Microarray analysis of *Pseudomonas aeruginosa* quorum-sensing regulons: effects of growth phase and environment. J Bacteriol 185:2080–2095

Walker TS et al (2004) *Pseudomonas aeruginosa*-plant root interactions. Pathogenicity, biofilm formation, and root exudation. Plant Physiol 134:320–331

Wei Q et al (2012) Global regulation of gene expression by OxyR in an important human opportunistic pathogen. Nucleic Acids Res 40:4320–4333

Weinel C, Nelson KE, Tummler B (2002) Global features of the *Pseudomonas putida* KT2440 genome sequence. Environ Microbiol 4:809–818

Wilderman PJ et al (2004) Identification of tandem duplicate regulatory small RNAs in *Pseudomonas aeruginosa* involved in iron homeostasis. Proc Natl Acad Sci U S A 101: 9792–9797

Williams P, Camara M (2009) Quorum sensing and environmental adaptation in *Pseudomonas aeruginosa*: a tale of regulatory networks and multifunctional signal molecules. Curr Opin Microbiol 12:182–191

Wyckoff EE, Lopreato GF, Tipton KA, Payne SM (2005) *Shigella dysenteriae* ShuS promotes utilization of heme as an iron source and protects against heme toxicity. J Bacteriol 187: 5658–5664

Yang B, Hoegy F, Mislin GL, Mesini PJ, Schalk IJ (2011) Terbium, a fluorescent probe for investigation of siderophore pyochelin interactions with its outer membrane transporter FptA. J Inorg Biochem 105:1293–1298

Yeterian E, Martin LW, Guillon L, Journet L, Lamont IL, Schalk IJ (2010) Synthesis of the siderophore pyoverdine in *Pseudomonas aeruginosa* involves a periplasmic maturation. Amino Acids 38:1447–1459

Youard ZA, Mislin GL, Majcherczyk PA, Schalk IJ, Reimmann C (2007) *Pseudomonas fluorescens* CHA0 produces enantio-pyochelin, the optical antipode of the *Pseudomonas aeruginosa* siderophore pyochelin. J Biol Chem 282:33553–35546

Youard ZA, Reimmann C (2010) Stereospecific recognition of pyochelin and enantio-pyochelin by the PchR proteins in fluorescent pseudomonads. Microbiol 156:1772–1782

Youard ZA, Wenner N, Reimmann C (2011) Iron acquisition with the natural siderophore enantiomers pyochelin and enantio-pyochelin in *Pseudomonas* species. Biometals 24: 513–522

Zheng P, Sun J, Geffers R, Zeng AP (2007) Functional characterization of the gene PA2384 in large-scale gene regulation in response to iron starvation in *Pseudomonas aeruginosa*. J Biotechnol 132:342–352